I0478506

# *The Bighorn Basin, Wyoming:*
## *Faults, Folds, and the Petroleum Trap*

*By*

*William A. Szary*

*Earth2Energy Educational Publishing*
*Copyright 2017. All rights Reserved.*

Copyright 2017, **William A. Szary**
Publication Date: November, 2016

Library of Congress Cataloging in Publication Data

Szary, William A.
The Bighorn Basin, Wyoming: Faults, Folds, and the Petroleum Trap

Includes: References & Index

1.  Geology-Wyoming; 2. Petroleum Geology-Wyoming; 3. Bighorn Basin
    Geology; 4. Bighorn Basin Petroleum Provinces; 4. Wyoming Tectonics

ISBN-13: 978-1540641342
ISBN-10: 1540641341

# Contents

Chapter 1. Western Regional US Geologic History      3

Chapter 2. The Bighorn Basin Stratigraphy and Sedimentation      27
    Paleozoic Sedimentation
    Mesozoic Sedimentation
     The Western Interior Seaway
    Cenozoic Sedimentation

Chapter 3. The Bighorn Basin Geologic Setting      53
    The Petroleum Trap
    The Basin marginal Anticlinal Trap

Chapter 4. Basic Petroleum Geology: Faults, Folds and Traps      58
    Folded Traps
    Fold Pattern Shifting
    Faulted Traps
    Stratigraphic Traps
    Carbonate Rock Facies
    Fluid Traps
    Combination Traps

Chapter 5. Bighorn Basin Geologic Structure      81
    Tectonic Events
     Bighorn Basin Tectonics
     Sevier Tectonics
     Laramide Tectonics
    Western Basin Marginal Folds
    Southern Basin Marginal Folds

Chapter 6. Basin Petroleum Province Play Zones      100
    Conventional Plays
    Hypothetical Plays
    Unconventional Plays

Index
References

# Chapter 1.

## Regional Western US Geologic History

The Western US was built out by shallow marine sea sediments, formation of spreading centers offshore, colliding island arcs, and retreating sea levels. Thrust faults developed when island arcs collided with the margin, compressing older rocks east, pushing rocks on top of younger rocks further inland. Island arcs continued to pile up against the margin. These collisions caused subduction trenches to jump west as each island arc crashed into the western margin. Volcanic chains developed when oceanic crust was pulled beneath the margin.

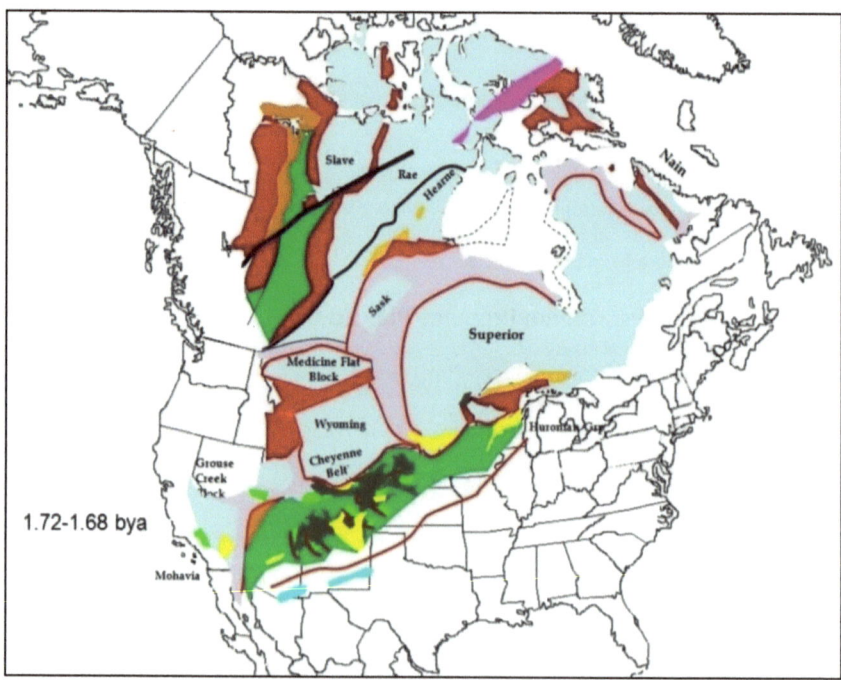

*Figure 1. The North American Craton was continuing to assemble between the Early and Middle Precambrian Era.*

Precambrian gneiss and granite formed basement complexes beneath Wyoming. The Cheyenne Belt was uplifted along the southern part of the state. Plutonic rocks were injected into basement rocks along the north and western part of the "Wyoming plate".

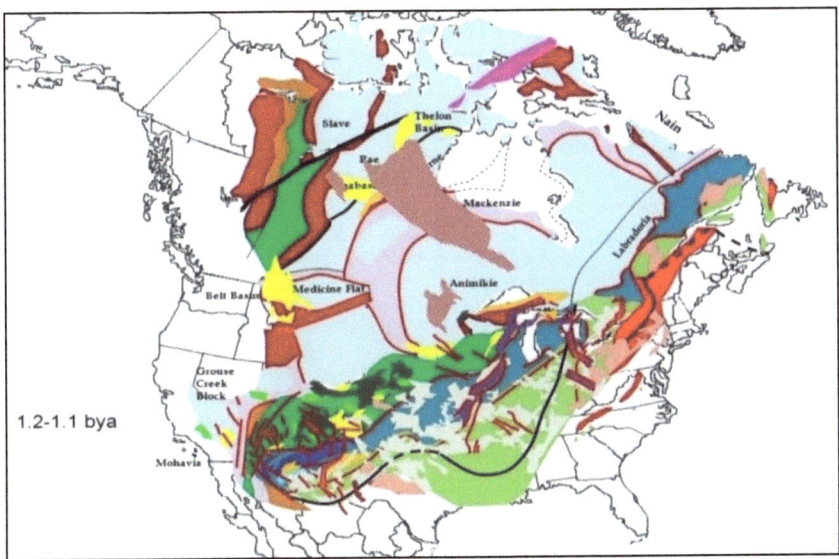

*Figure 2. North America attached to Rodinia, as igneous intrusions invaded mountain belts, and flood basalts erupted in Canada. The southwest basin and range faults developed at the same time.*

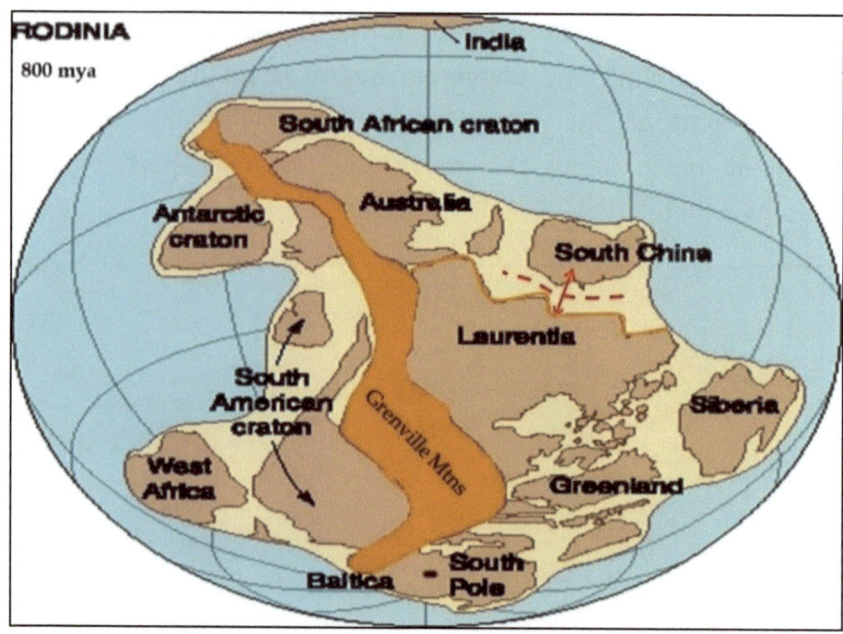

*Figure 3. The Rodinian supercontinent assembly included the Grenville Mountains and a rift zone separating South China from Laurentia.*

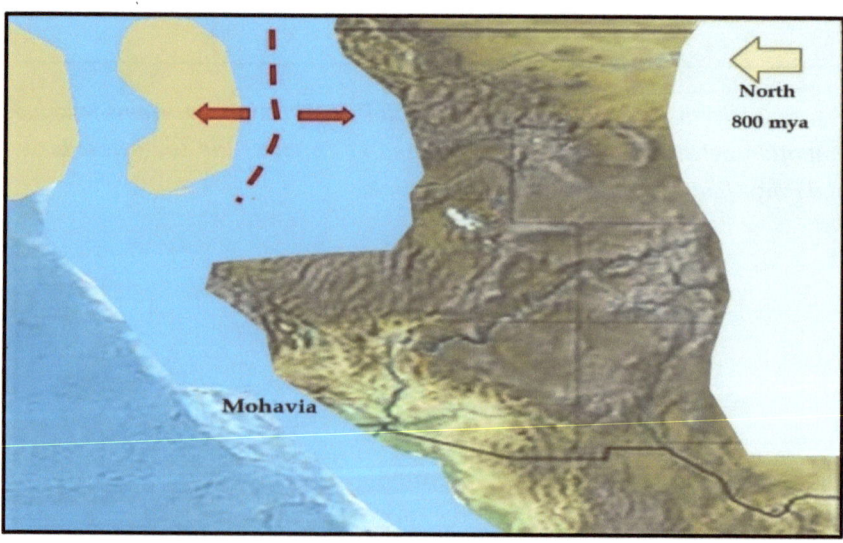

*Figure 4. Continental margin rifting occurred along the western United States, when seas invaded the continental margins of California, Washington, and Oregon. Seas extended into western Utah, towards Late Precambrian time. Mohavia was the precursor terrane to the Mohave Desert Region of California.*

6

About 1.7 billion years ago, the Western US may have resembled something like modern SE Asia (**Figure 1**). Various tectonic blocks accreted together, uplifting mountain belts along suture zones.

Around 1.1 billion years ago, North America was attached to Rodinia (**Figure 2**). The SW part was positioned adjacent to Australia and/or Antarctica, or Siberia. A volcanic arc complex was positioned off southern North America. Sedimentary rocks formed along the North American passive margins.

About 800 million years ago, Laurentia was positioned around 45 degrees south latitude (**Figure 3**). The west coast faced north and the east coast faced south. The northern Laurentian margin was positioned somewhere in eastern Idaho, central Nevada, and central California at this time. A rift opened up along a mid-oceanic ridge in eastern Washington & Oregon. South China was separated by a rift from northern Laurentia. South China was drifting towards the Arctic. India was located in the Arctic Circle.

The Grenville Mountains occupied the southern Laurentian margin extending northwest through Australia and South Africa. The mountain range separated South America from Baltica.

The western US continental margin was positioned east of a mid-oceanic spreading ridge (**Figure 4**). The ridge was positioned along the Idaho-Montana state line extending south through central Nevada into south central California, west of Mohavia. The rift zone was positioned in eastern Washington and Oregon. Where the rift was positioned south of Mohavia is unknown due to destruction by later tectonic activity in Nevada and California.

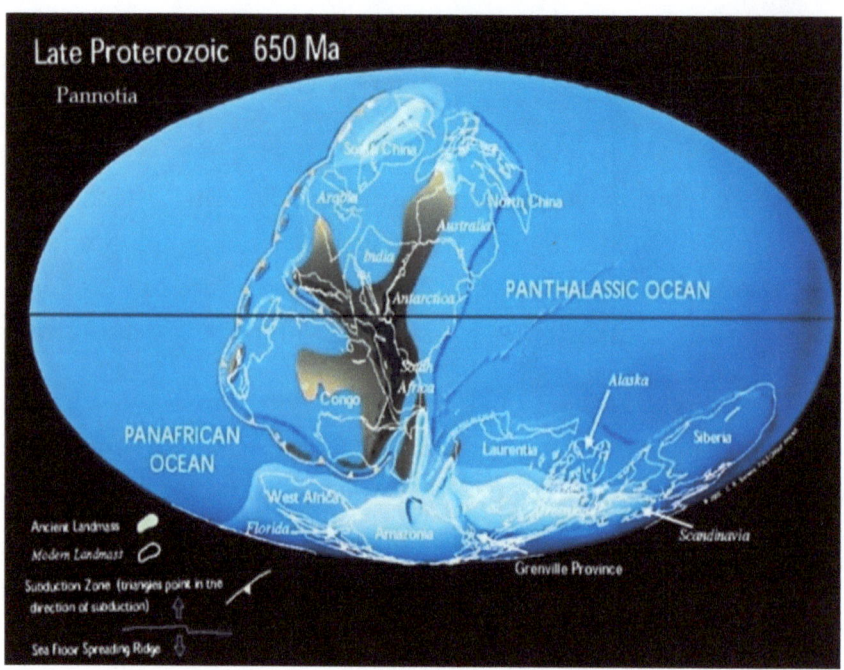

*Figure 5. Pannotia formed as a linear supercontinent after Rodinia broke apart. Portions of Australia, India, Antarctica, Arabia, and Africa made up the continent.*

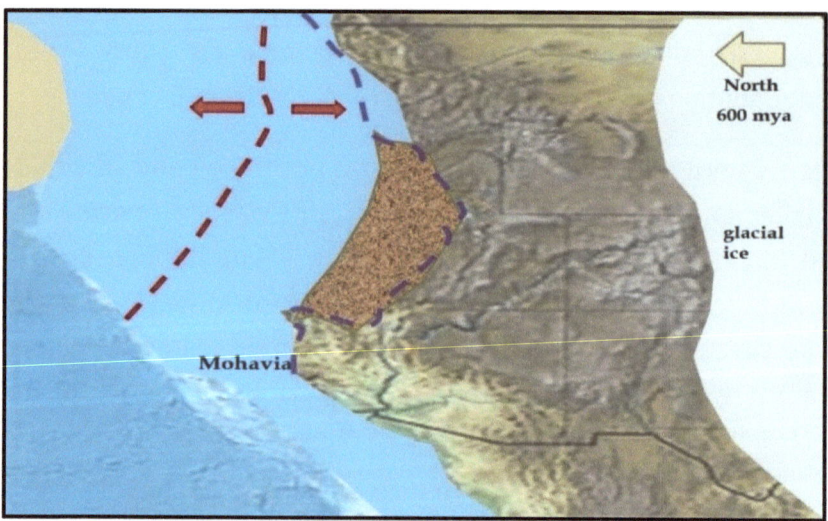

*Figure 6. Rifting of the western US margin continued along the western Laurentian margin. Sediments were shed from the coastal margin onto the continental shelf (brown stipple). The purple dashed line represents the continental margin.*

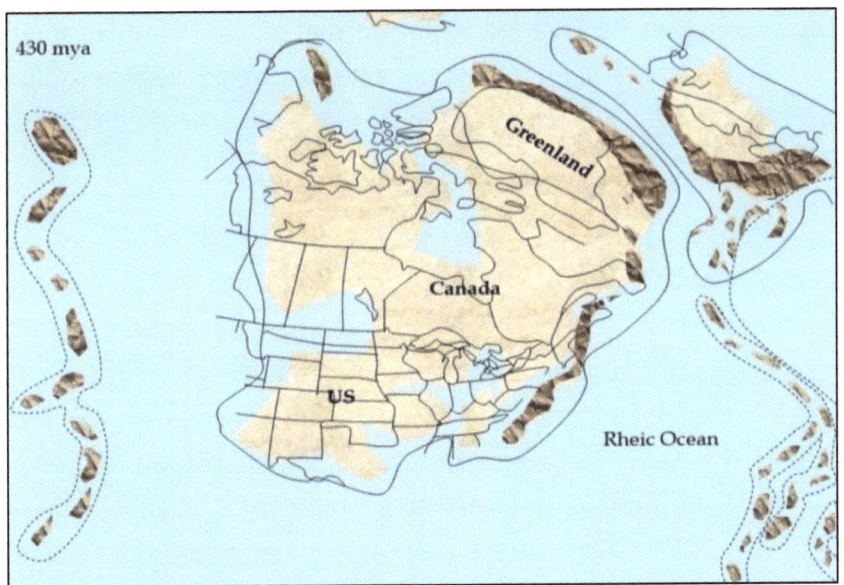

*Figure 7. Shallow seas invaded Laurentia. Island arc systems approached from the east and west directions. Western Wyoming was submerged during Ordovician time. Eastern Wyoming remained emergent.*

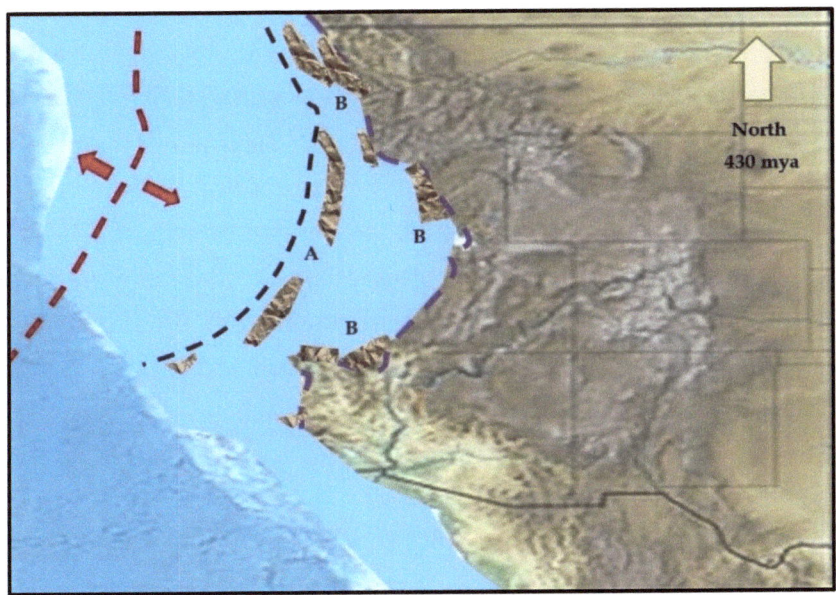

*Figure 8. The Antler Arc formed as an island arc approached the western margin from the Panthalassic Ocean. As the arc approached the continent, seas began to withdraw. The letters mark the positions of the arc as it collided with the margin.*

The breakup of Rodinia led to the formation of Pannotia, about 650 million years ago. Laurentia remained at 45 degrees south latitude with most of the eastern portions covered by glacial ice (**Figure 5**).

Between 700 and 600 million years ago, the continental margin shifted inland (**Figure 6**). Rifting thinned and stretched the margin. A deepening sedimentary basin formed along the western passive margin. Thinning interior sediments accumulated on the shelf, and thickening sediments accumulated in the deeper offshore area to the west. Sandstone, mudstone, and limestone formed in the shallow marine sea off the continental margin. These sediments formed the future mountainous margin. Rifted micro-continental fragments remained off the SW North American margin during the Paleozoic Era. Several of these micro-continents later accreted onto Mexico and California during the Mesozoic Era.

About 430 million years ago during the Silurian Period, Laurentia drifted north into the Equatorial region (**Figure 7**). Shallow seas occupied the eastern and western interior. Between 430 and 395 million years ago, deposition continued along the western North American margin and a volcanic arc began to approach the continent from the west. The Antler Island Arc became wedged beneath the marginal subduction trench (**Figure 8**). During the Late Devonian Period, the approaching arc collided with the North American passive margin. Collisions between the arc and continent fragmented the arc system.

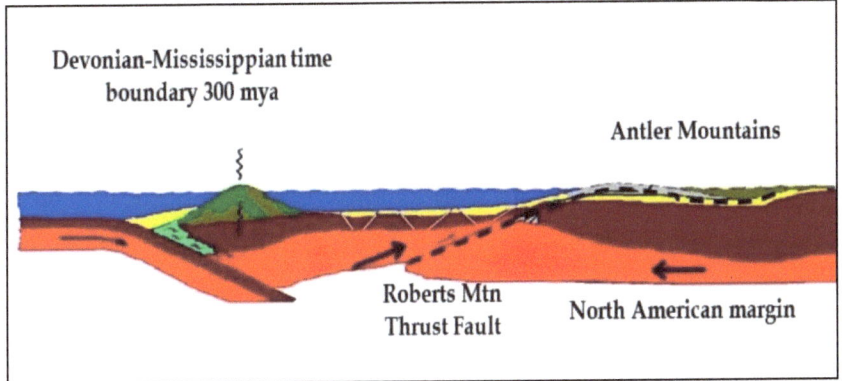

*Figure 9. Antler Mountain thrust faulting formed after the Antler Arc collided with North America. The approaching new arc began to push oceanic crust and mantle material on top of the Antler Mountains forming the Roberts Mountain Thrust Fault.*

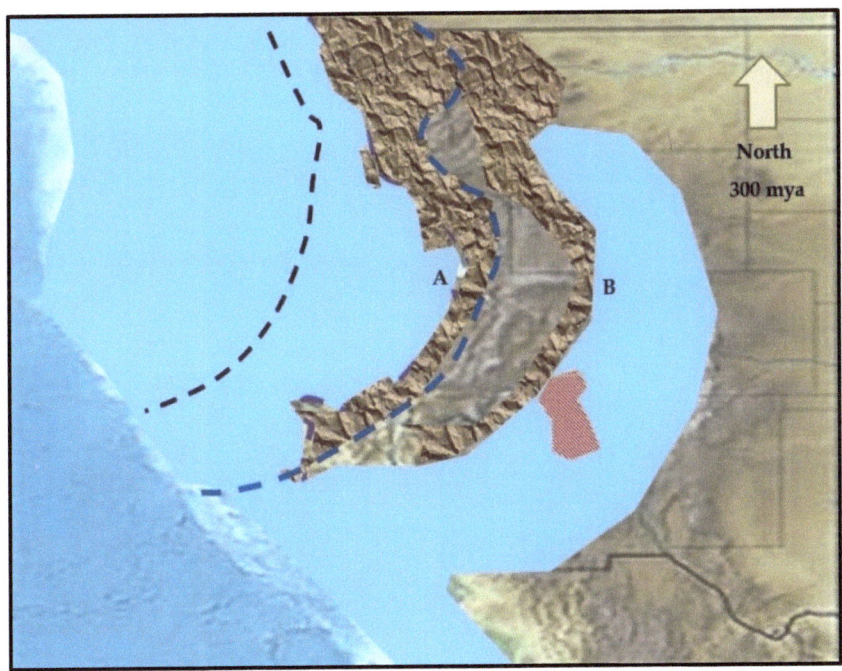

*Figure 10. Antler fore arc thrusting continued to push the colliding island arc system eastward (A to B) forcing the interior seas to retreat from the western U.S.*

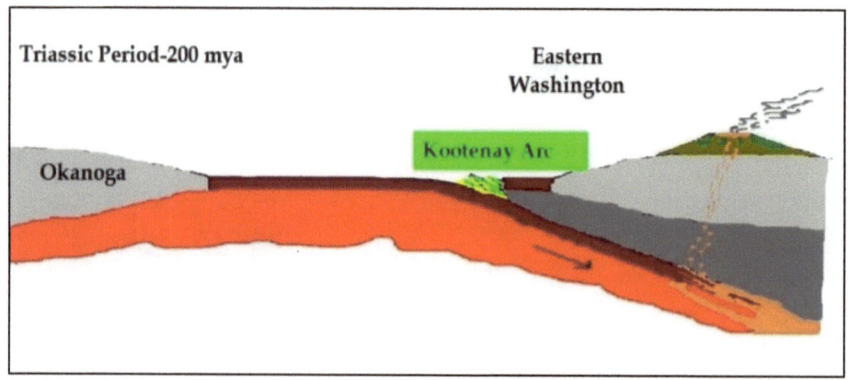

*Figure 11. The Kootenay-Okanogan Arcs collided with eastern Washington. The first arc to collide was the Kootenay Arc, resulting in the creation of a continental volcanic arc from the subducting oceanic slab. The Okanogan Arc was approaching, positioned offshore at this time.*

*Figure 12. Subduction zones jumped west. A series of island arc collisions accreted onto the western US. One of the arc systems, called the Okanogan terrane, collided with North America. Dashed red lines show the approximate position of former subduction trenches.*

Approximately 300 million years ago from the Late Devonian into the Early Mississippian Period, the Antler Mountains were pushed on top of the western continental margin along a thrust fault located near eastern Washington and Oregon (**Figure 9**). The position of the subduction zone was located further west. The continental margin positioned in western Montana and southeastern Idaho was deformed by later faulting. When the Rocky Mountains uplifted, the Antler Mountains were eroded and buried by sediments shed from the Rockies.

The Antler fore-arc was thrust eastward on top of the Paleozoic sinking margin, forming the Roberts Mountain Thrust Fault. During thrusting and collision, accretion between the Antler arc and margin formed low highlands. Rocks were piled on top of the sinking margin, adding weight to the continental margin as subsidence increased. Combined sedimentation and uplift formed the Antler foreland basin.

The foreland basin was filled with marine sediments. A slight bulging of the stable interior occurred east of the foreland, in the four corners region (NM, CO, UT, AZ). When compression ended, uplifting ceased. Subsidence of the passive margin resumed (**Figure 10**).

About 200 million years ago, during the Triassic Period, the supercontinent Pangea began to break up along mid oceanic ridge systems. The Atlantic Ocean began to form when the African and Europe Plates began to move away from the North American Plate. The rifting of the Atlantic Ocean basin caused the Pacific Plate to begin sliding beneath the western North American continental margin. Oceanic crust and sediments begin to scrape off against the western continental margin. These sediments formed the Kootenay Arc, located in eastern Washington. The Okanogan subcontinent was positioned off shore from the Kootenay arc subduction zone (**Figure 11**).

Seas retreated again towards the west. The Late Triassic subduction zone jumped off the west coast margin, again. The Kootenay Volcanic Arc wedged onto the eastern Washington margin shortly before the Sonomian arc collided with the southwest. The arc position most likely trended through central Oregon beneath younger volcanics, possibly linking up with the California Sierra Mountains (**Figure 12**).

*Figure 13. Subduction trenches continued forming off the western North American margin when Pangea broke apart. The Mexican craton began forming when fragmented land masses and island arcs continued approaching from the west.*

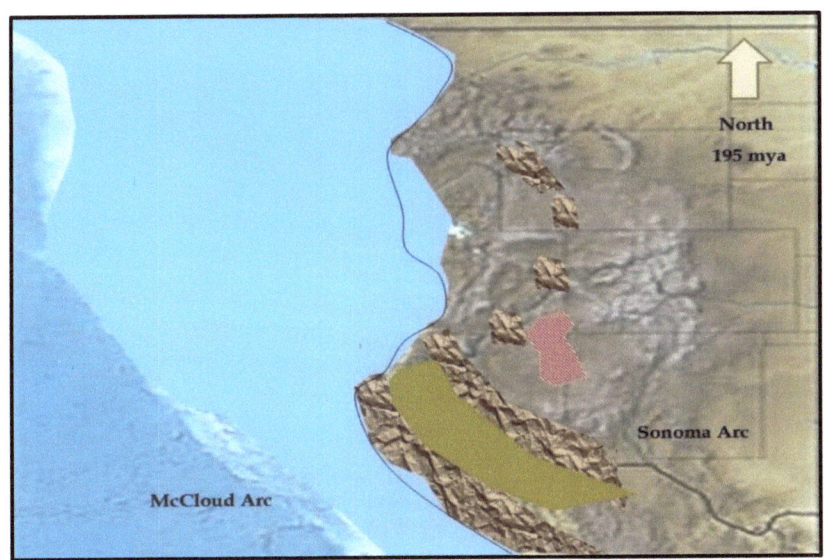

*Figure 14. Island arc collisions, thrust faulting, and mountain building continued along the west margin, building out the continent into a series of continental volcanic arcs, forming the western US. The McCloud Arc collided with the Sonoma Arc in the southwest.*

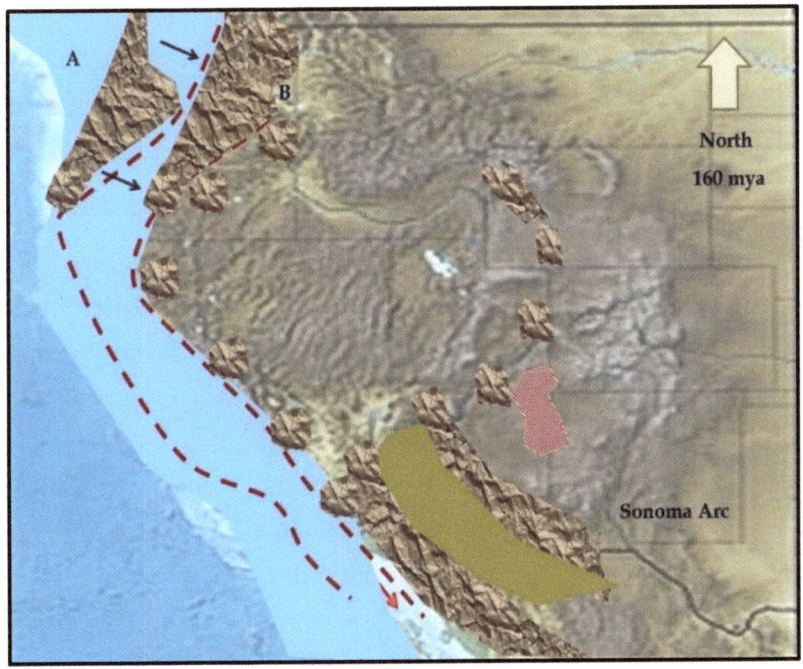

*Figure 15. The Okanogan micro-continent collision forced the Pacific trench to jump west (dashed lines). Letters A and B show the approaching arc.*

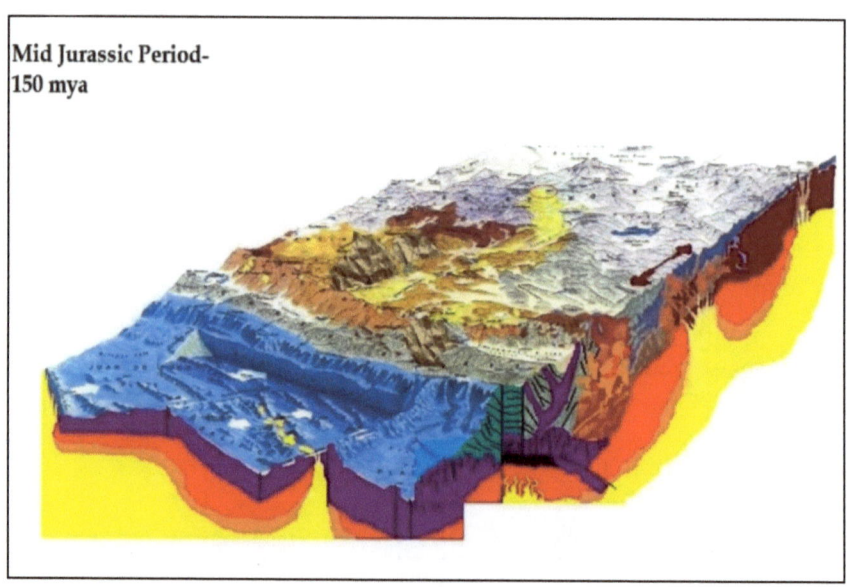

Mid Jurassic Period- 150 mya

*Figure 16. Accretion & mountain uplift continued to build out the western continental U.S. The block model shows the various accreted terranes represented by the different colors (see text for explanation).*

Plate motions shifted between the ancestral Pacific Plate and western North American margin. An eastward dipping subduction zone developed along the western Gondwana margin, south of California (**Figure 13**).

The inland sea retreated during island arc collision, uplift, and built out of the western margin. The western continental margin was positioned along eastern Idaho, extending south. The SW margin was cut off by left lateral shearing. Lateral motion means horizontal motion with the far side of the fault moving to the left side. Shearing means one block slides past another, warping the region positioned between the two blocks. Various arc collisions formed the Sonomian mountain belt. The McCloud arc became part of the Sonomia arc when it collided with North America. The back arc basin was thrust eastward over the western margin of North America.

The Cache Creek inter-arc basin was trapped between the McCloud arc and Sonomia arc blocks, forming an attached wedge to the continental margin.

By Late Triassic time, the southwest was the location of a new mountain volcanic arc system, as a back arc basin developed between the arc and stable interior on top of the former Antler and Sonomian mountains. When the Sonomian arc collided with the stable Arizona interior, the pushing and pulling of the Pacific Plate against the stable continental interior caused the Arizona interior to form thrust and normal faults during the arc- continent collision, between 190 and 136 million years ago. Behind the arc, a high dune field formed, mixing with arc volcanics throughout AZ, CA, and NV (**Figure 14**).

With the last continental accretion occurring (Foothills terrane) against the western slopes of the Sierra Nevada Mountains, 160 million years ago, the Sierra trench jumped westward to the Coast Range.

The Okanogan subcontinent docked against the Kootenay arc, causing the trench to jump westward into central Washington, south through Oregon and through the Californian Coast Ranges (**Figure 15**).

About 150 million years ago, the west was going through a transition period (**Figure 16**). The subduction trench was beginning to shift from the Washington-Idaho and Oregon-Idaho boundary into central Oregon. It is believed trench shifting did not begin in Washington until the Okanogan micro-continent docked in eastern Washington. In eastern Orégon, the Wallowa Mountains accreted to the western continental margin. The Blue Mountains accreted onto the central Oregon margin in an event called the Wrangellia Orogeny. The Klamath Mountains were already accreted onto southern Oregon and northern California.

The California trench jumped from the western Sierra Mountains to the Coast Ranges. The Sierra Mountains were formed in central California.

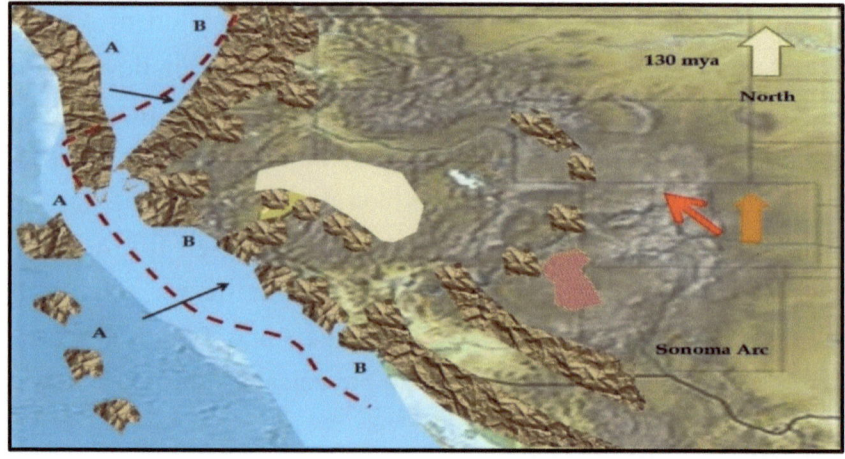

*Figure 17. It is believed Wrangellia was drifting north throughout the Paleozoic from southern latitudes. During the Mesozoic Era, Wrangellia collided with western North America, uplifting the Blue Mountains of Oregon and the central Nevada mountain region. Red and orange arrows indicate changes in plate motion from north to northwest.*

*Figure 18. Island arc collisions continued. North America drifted westward and the oceanic plate pushed eastward (red arrows). Volcanic chains began erupting from the arc collisions. Subduction remained active along the margin. Trenches continued to jump westward.*

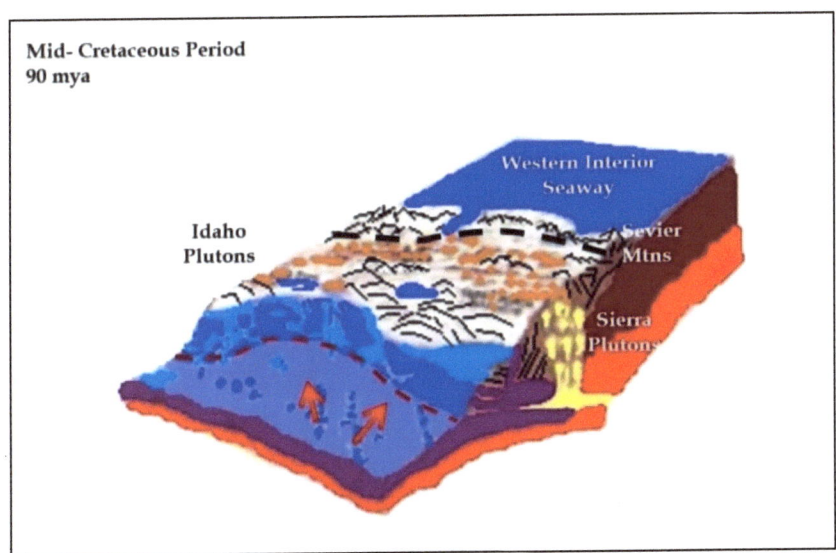

*Figure 19. Sevier Mountain uplift began and the Western Interior Seaway flooded the interior parts of North America. The Sierra Nevada plutons were intruded at this time, forming the basement for the mountain range. The Idaho Batholith also formed during Kula Plate subduction beneath the margin.*

*Figure 20. Baja British Columbia was positioned in Mexico when the East Pacific Rise formed off the western coast.*

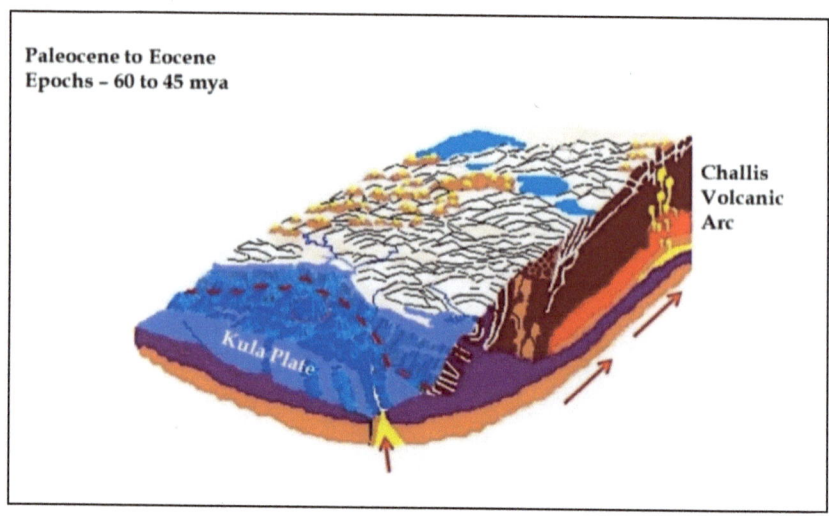

*Figure 21. The Kula Plate was being consumed beneath the western margin, giving way to the Farallon plate. The Challis Arc began erupting in Idaho. The Western Interior Seaway receded.*

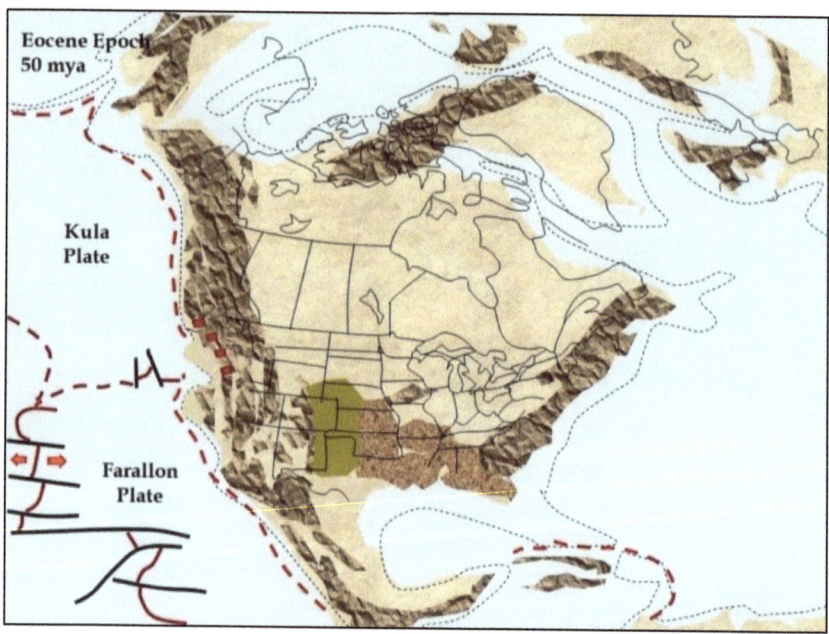

*Figure 22. The Laramide Mountains formed during continued Farallon Plate subduction beneath the western margin. Accreted terranes were pushed on top of each other towards the interior. The Mendocino Fracture Zone formed as one of many transform faults offsetting displacement along the East Pacific Rise.*

Continued subduction along the SW margin caused plateau uplift in Nevada during the Jurassic Period (**Figure 17**). A large oceanic plateau, called Wrangellia, collided with the Pacific Northwest margin, migrating northeast until it contacted the margin. Wrangellia continued attaching onto the margin causing older accreted crust to slice apart along horizontal faults, including fragments of Wrangellia. Some fragments were thrust into Nevada, starting the Nevadan orogeny. By Late Jurassic time, the Nevadan orogeny was fully active. At the same time, terrane suturing, thrusting, lateral faulting, and rifting was occurring throughout western North America. The North American plate shifted direction to the northwest. The Nevadan arc ceased eruption. Broad basins developed on the landward side of the arc. The western interior began to drain by major rivers flowing to the northeast.

Small island continents about the size of Borneo, Japan, New Zealand, and New Guinea began docking with the western North American continent about 100 million years ago (**Figure 18**). The first collision was the Okanogan micro-continent in northeastern Washington. The attachment of Okanogan to the North American Plate closed off the Kootenay trench. The Pacific Plate motion was directed east and the North American Plate motion was directed west. Changing plate motions resulted in a new trench and subduction zone jumping to the west of Okanogan. A new volcanic arc formed east of the trench. Younger volcanics buried evidence of the arc position in Oregon.

About 90 million years ago, the Kula Plate subducted beneath the Oregon-California coast line (**Figure 19**). Partial melting of the oceanic plate beneath the continent resulted in molten magma rising into the lower crust, reaching the surface as a volcanic arc. This arc formed the Idaho and Sierra Batholiths.

The arc shifted eastward when the paleo-Pacific plate moved to the northeast, ending the Sevier orogeny. Thrust faulting pushed continental crust eastward through Utah and western Wyoming during the Sevier Mountain uplift. Erosion of the Sevier Mountains supplied sediments to the Western Interior Seaway.

The rising mountains blocked off a shallow inland sea which occupied portions of Montana and Wyoming, extending to the Gulf of Mexico. East of the subduction zone, the oceanic plate moved north against the continental plate along horizontal slip faults. Accreted terranes were shifted northward.

Baja British Columbia (BC) was positioned in the Baja California region during the end of the Cretaceous Period (**Figure 20**). Baja British Columbia was transferred north along a right lateral fault until it reached its present position on the Pacific NW coast. The eastern Mesozoic arc was extinguished during the transport of Baja BC.

Between 60 and 45 million years ago, the Kula Plate flattened out, continuing subducting beneath the Oregon-Northern California coasts, reaching far inland beneath Idaho, Montana, and Wyoming (**Figure 21**). A new spreading center opened up, separating the Kula Plate from the Farallon Plate. Seamounts, carried along on the surface of the subducting ocean plate, collided with the northeastern Oregon coast. The Sierra Mountains eroded down to low lying hillsides. Further inland, in central Idaho, the Challis Volcanic Arc (known today as the Absaroka Mountains) formed from partial melting of the Kula Plate beneath the continental crust. Molten magma, rising to the surface, formed a volcanic chain in central Idaho.

Idaho's rising volcanic chain began pushing up mountain belts in Montana and Wyoming with molten magma raising the Bighorn, Grand Tetons, and other interior mountain belts when the subducting Kula Plate continued sinking beneath the inland crust. During uplift of the interior mountains, the inland sea receded into large swamps and vast lakes.

During the Eocene Epoch, 50 million years ago, the Farallon Plate rapidly subducted beneath the western margin. The East Pacific Rise began to approach North America (**Figure 22**).

The Laramide Orogeny was actively rising in the central western interior, and the western Rockies remained elevated. Sediments shed from the rising Laramie Mountains covered the Great Plains. Magmatism migrated west. A new arc formed along the marginal mountain belt and the foreland basins began filling along the Pacific margin.

The East Pacific Rise closed in on North America. The Mendocino Fracture Zone formed and collided with the continental margin. The triple junction formed when the Farallon Plate broke apart.

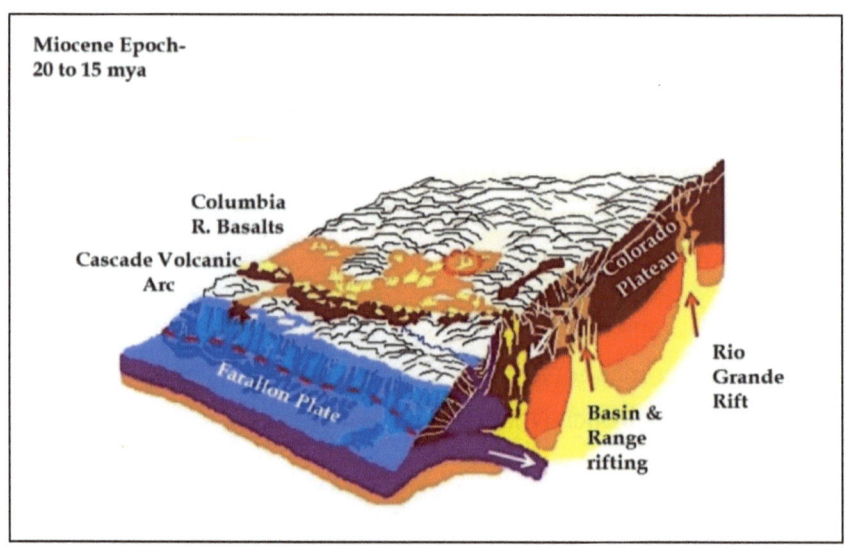

*Figure 23. The Cascades Volcanic Chain formed along the western margin when the Farallon Plate continued to dive beneath the margin. Basin & Range faults began to uplift mountains in the interior western region, tugged by the descending Farallon Plate which pulled down the continental crust. The Colorado Plateau uplifted when the Rio Grande Rift began to pull apart the eastern edge of the western accretionary belts.*

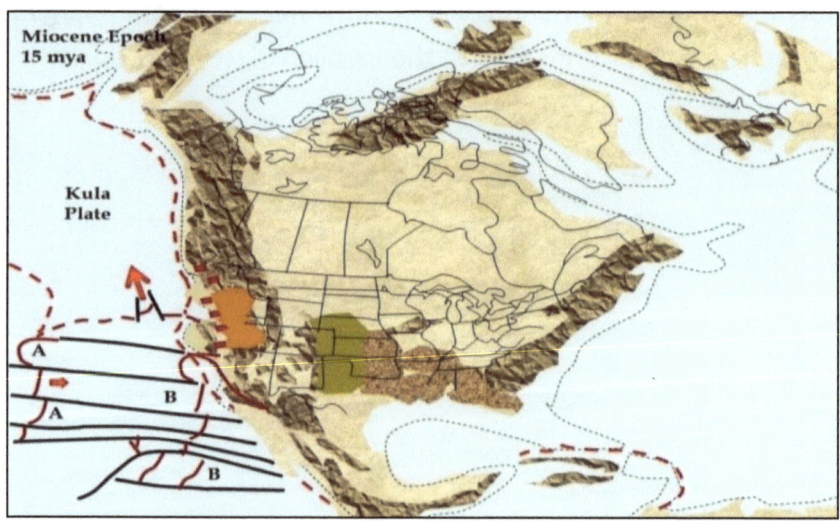

*Figure 24. The Mendocino Fracture zone collided with western North America. The continental margin began to break apart along faults. The Cascade Arc continued erupting and Kula Plate motion changed direction to the northwest.*

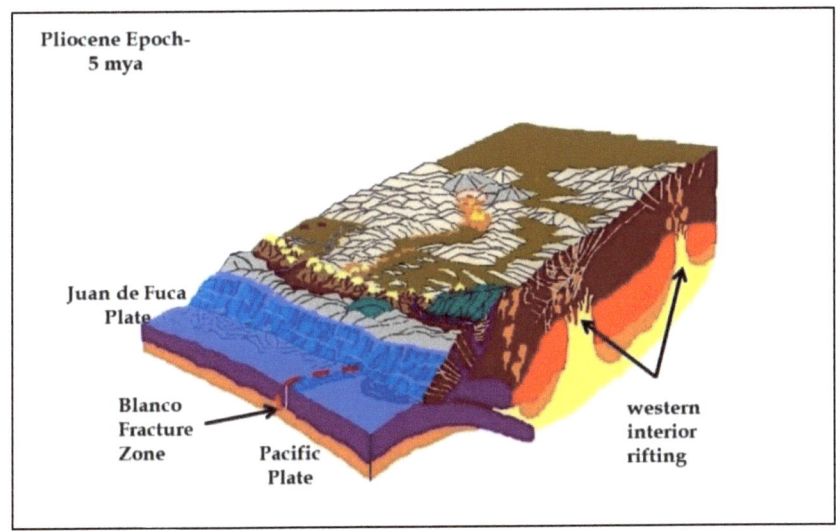

*Figure 25. The Blanco Fracture Zone separated the Juan de Fuca plate from the Pacific plate. San Andreas faulting and Basin & Range faulting began. The Yellowstone Hot Spot was activated.*

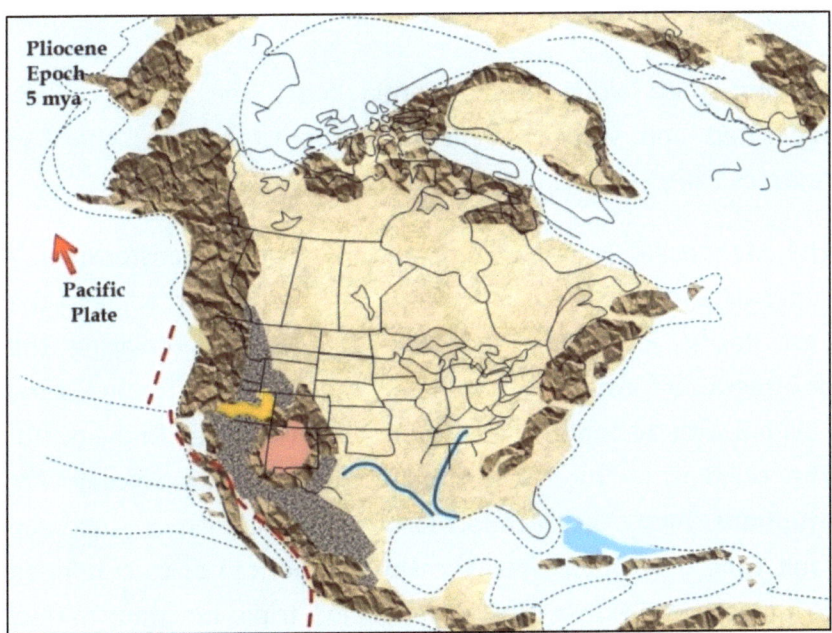

*Figure 26. The Olympic Mountains uplifted during the build out of Washington. The Snake River Plain began erupting as an extension of the Yellowstone volcano (pink). The Kula Plate was entirely consumed beneath the Aleutian Islands of Alaska when the Pacific Plate took its place.*

Between 20 and 15 million years ago, the descending Farallon Plate steepened, slowing down from 15 centimeters per year to 2 cm per year (**Figure 24**). The Cascade Volcanic Arc erupted in Washington, Oregon, and northern California. The Basin and Range began to rift apart, tugged by the pulling and thinning of the continental interior crust in opposite directions by the Farallon Plate and by resistance from the eastern continental crust. Rifting formed the Rio Grande Rift in New Mexico and Colorado, raising the Colorado Plateau. The plateau was wedged between the Basin & Range-Rio Grande Rift zones.

The Columbia River Basalt flows spilled out onto the eastern Washington-Oregon landscape, and the Yellowstone Hot Spot erupted along the Oregon-Idaho border when the continent was pushed over a shallow part of the mantle.

Further south along the California coast, the Farallon Plate separated and wedged beneath the crust forming the San Andreas Fault.

The Mendocino Fracture Zone, including other transform fault zones extended to the east towards the margin. Spreading of the East Pacific Rise continued towards the east replacing the disintegrated former Farallon Plate (**Figure 25**). The fracture zone collided with the margin when subduction continued to the north. The colliding fracture zone initiated the Cascade Volcanic Arc eruptions. Along the Mendocino transform margin, a strike slip fault zone developed when the North American plate continued to move west. North of the Mendocino triple junction, normal subduction continued. The Cascade arc remained active. To the south of the junction, crustal extension spread north and ended to the south. Volcanism covered wide areas of Washington and Oregon. Basins received thick lava flow and ash deposits.

The Gulf of California opened. A large block of the southwestern continental margin slid north on the Pacific Plate along the San Andreas Fault.

The Blanco Fracture Zone broke apart the Juan de Fuca Plate from the Pacific Plate (**Figure 26**). San Andreas Faulting expanded into northern California. Cascade Arc eruptions continued into northern Washington and British Columbia. Basin and Range faulting began when the Juan de Fuca plate and the San Andreas Fault began tugging at the central western interior crust, rifting the crust apart. The Sierra and Klamath Mountains were being pushed westward by the Basin & Range rift. Yellowstone Hot Spot eruptions produced Snake River Plain lava flows.

When the interior of the continent dried out, the Great Plains region developed above the stable continental craton. Most of the Western US was completely assembled by the beginning of the Pliocene Epoch (**Figure 27**). Oceanic plate sediments continued to scrape off onto the Washington coastline, uplifting the Olympic Mountains. The Snake River Plain continued erupting, and the North American Plate slid over a hot spot located in southeastern Idaho. The San Andreas Fault connected the Gulf of California to the Mendocino Fracture Zone, taking up slip by motion occurring along the Pacific Plate in the same direction. Basin and Range Faulting continued. San Andreas Fault began motion to the northwest.

The Colorado Plateau was raised to its peak elevation. Interior drainage began dissecting the surface. The Mississippi River began developing drainage within the mid continental region.

# Chapter 2.

## The Bighorn Basin:
## Stratigraphy & Sedimentation

### Paleozoic Sedimentation

During the Paleozoic Era, warm, shallow tropical seas covered most of Wyoming along a passive margin developed along the western North American continent. Wyoming was positioned near the Equator. Seas advanced and retreated across the state many times when the state was part of a regional shelf margin covering the western continental margin from most of North America into the Yukon territories of Canada. Sandstone, shale, and limestone were deposited under these conditions.

During Cambrian time (570 to 505 million years ago), shallow seas occupied most of the central part of Wyoming. Sand and mud accumulated on top of a continental shelf where sandstone, shale, and carbonate rich limestone formed. Evaporating seas concentrated magnesium brine resulting in dolomite sedimentation. Corals, crinoids, brachiopods, and bryozoans accumulated during this time. Death Canyon Limestone and Park Shale represented warm tropical seas at a time when North America was closer to the Equator. Flathead Sandstone represented shorelines adjacent to an eroding continental margin (**Figure 27**).

Seas partially retreated from the eastern part of Wyoming leaving behind an unconformity in the southeastern half of the state during the Ordovician Period. A stable carbonate platform was present in the western part of the state, expanding into central and northeastern Wyoming. Partial evaporation changed limestone into Bighorn dolomite when calcium and magnesium brines began to accumulate.

Minor crustal movements allowed for periods of erosion and non-deposition to remove sediments from the stratigraphic record (**Figure 28**). Silurian sediments were eroded from the state during the collision between the Antler Arc and the western continental margin.

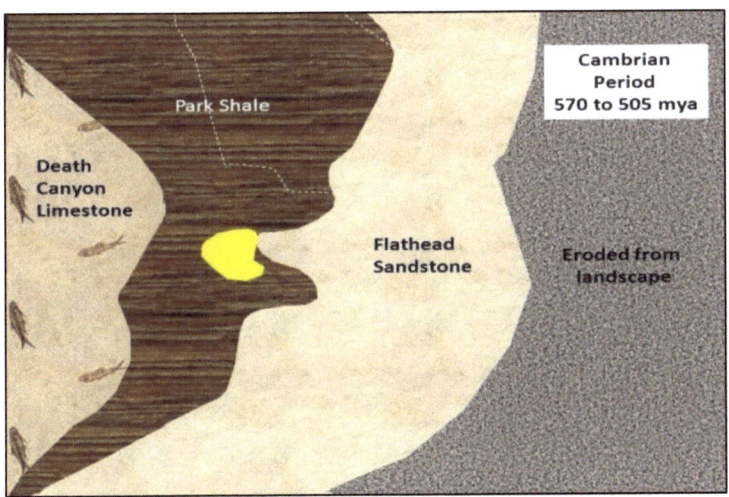

*Figure 27. Marine sedimentary rocks deposited during the Cambrian Period covered Wyoming (Source: Redrawn from Lageson and Spearing, 1988).*

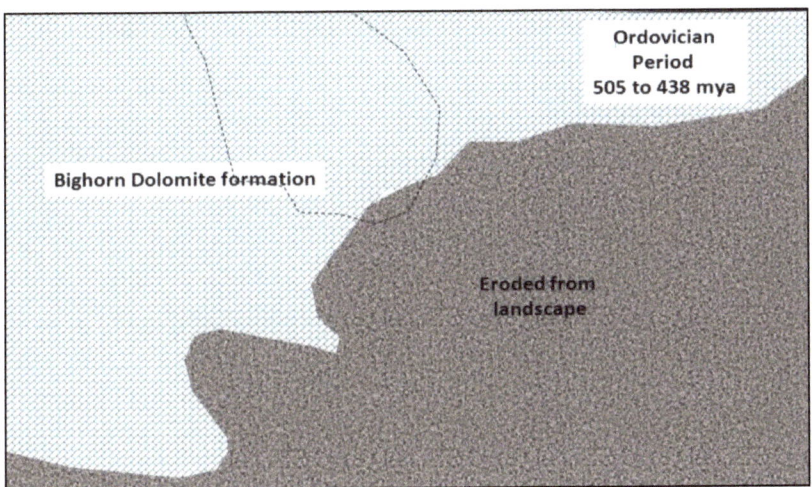

*Figure 28. Ordovician dolomite covered most of the western, central, and northeastern parts of the state during evaporating sea retreat cycles. (Source: Redrawn from Lageson and Spearing, 1988).*

Seas continued withdrawing during the Devonian Period leaving behind a mixture of dolomite and limestone cover in the western half of Wyoming. The eroded landscape of central and eastern Wyoming expanded westward. Along the northern boundary with South Dakota, Jefferson Formation limestone was deposited In the central western part of the state. Darby Limestone was deposited in the central and southwestern regions (**Figure 29**).

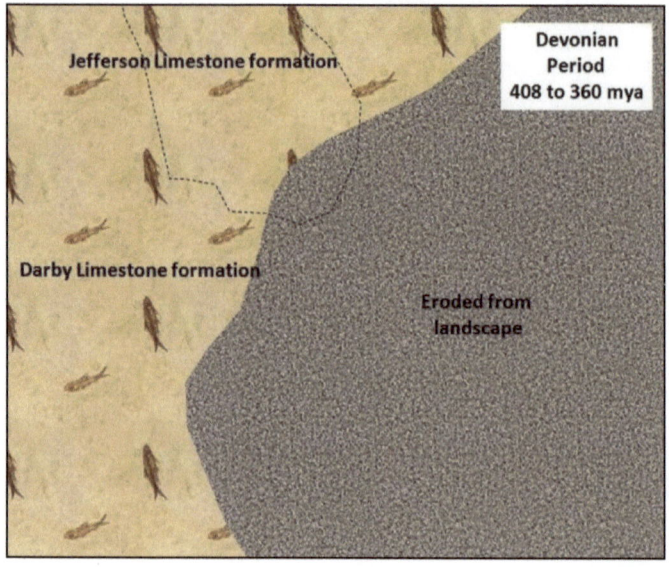

*Figure 29. Devonian erosion continued to expand westward during continued sea retreat cycles. Limestone began mixing in with dolomite in the western part of Wyoming. (Source: Redrawn from Lageson and Spearing, 1988).*

Around 300 million years ago, a set of island arcs collided with the North American margin when Europe separated from North American during the breakup of Pangea. These collisions formed the ancestral Rocky Mountains in Colorado.

The Antler Arc initially collided during the Devonian and Mississippian Periods, forcing the subduction trench to jump seaward. The new trench developed a second island arc which approached the continental margin.

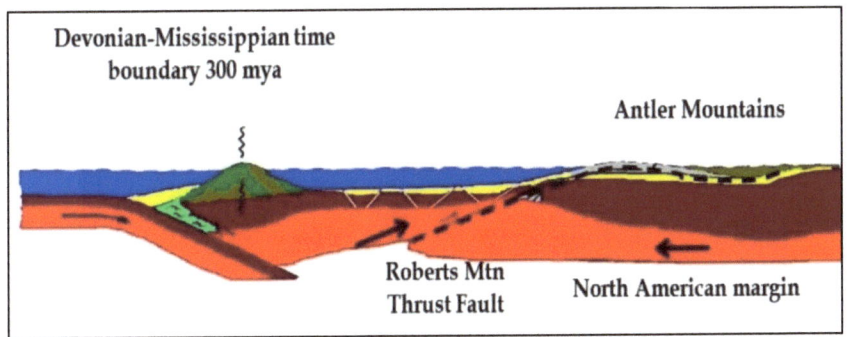

*Figure 30. Island arc collisions were responsible for developing the Antler Mountains, and Roberts Mountain Thrust Fault. (Source: Szary, 2014).*

Before the second arc collided, the arc pushed oceanic crust on top of the continental margin forming the Roberts Mountain Thrust Fault. The Antler Mountains Uplifted along with the Ancestral Rocky Mountains. Erosion, caused by the uplifting Rocky Mountains, washed off the eastern edge of the mountain front. More recent uplift along the Rocky Mountains tilted the Pennsylvanian-Permian Fountain Formation into angled rock exposures (**Figure 30**).

The Wyoming carbonate platform expanded during the Mississippian Period covering the entire region with the exception of local areas which either built up above the platform or were eroded by advancing seas resulting in erosion of sediments from the landscape. Guernsey Limestone formed in the eastern region, Madison Limestone formed in the western regions, and sandy limestone formed in the south central region where erosion of uplands contributed sand to the platform (**Figure 31**) between 360 and 320 million years ago.

Seas advanced, depositing Amsden Limestone which represented earlier erosion involving a period of widespread cave and karst development ending around 312 million years ago.

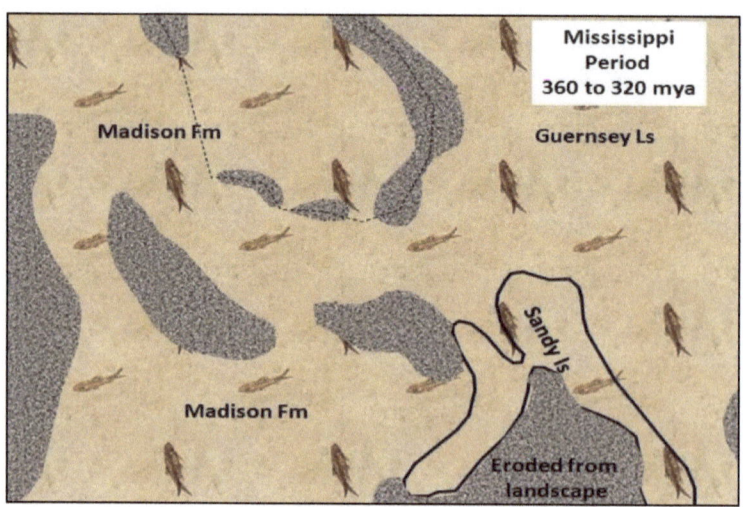

*Figure 31. Mississippian Period deposition covered the entire state with limestone and sandy limestone. Most notable, the Guernsey and Madison Limestone were deposited during this time. (Source: Redrawn from Lageson and Spearing, 1988).*

Tensleep sandstone covered the Madison Limestone during late Pennsylvanian time through a combination of marine and shoreline depositional conditions. Wind blown (eolian) deposits were part of the Tensleep depositional environment. Phosphoria limestone, sandstone, and black shale were deposited in a shallow seaway partially covering western Wyoming during infilling of the Antler foreland basin. Shallow water zones near deep, upwelling, nutrient rich marine water developed phosphorite deposits. Within the Phosphoria formation, Dryhead Agate developed under hydrothermal conditions within sedimentary host rock (**Figure 32**).

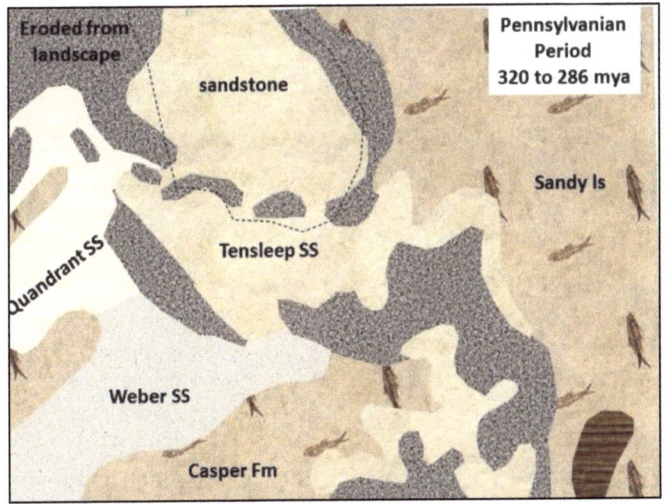

*Figure 32. The rising ancestral Rocky Mountains in Colorado influenced Wyoming geology. The northern prong of the ancestral Rockies called the Pathfinder Uplift extended into southeastern Wyoming. Sand deposits were shed onto the carbonate platform mixing in with limestone in the eastern part, covering the platform in the central and western parts of the platform. (Source: Redrawn from Lageson and Spearing, 1988).*

Seas began to evaporate from the carbonate platform in the eastern part of the state where Goose Egg Formation halite and anhydrite precipitated in shallow depressions on top of the platform within lagoons and tidal flats during the Permian Period. Sands began to invade limestone in the eastern central part of the platform. Deeper colder seas prevailed in the western part where upwelling occurred along the platform margin precipitated phosphate. Deeper siliceous sediments formed chert (**Figure 33**).

The beginning of the Mesozoic Era witnessed island arc collisions off the Western North American continental margin. Around 195 million years ago, British Columbia, Idaho, Western Utah, and Western Arizona uplifted from island arc collisions with the western margin. The Pacific Northwest region was submerged by the Pacific Ocean at this time.

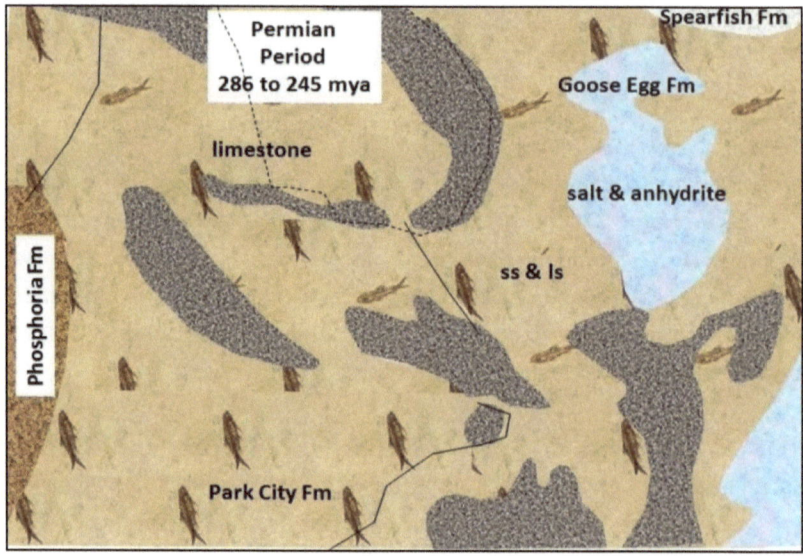

*Figure 33. Spearfish Formation red sands and siltstone formed in the northeast corner of the state while evaporites precipitated in evaporating seas from shallow depressions formed on top of the carbonate platform. Areas of non-deposition occurred where the platform was exposed to subaerial erosion. (Source: Redrawn from Lageson and Spearing, 1988).*

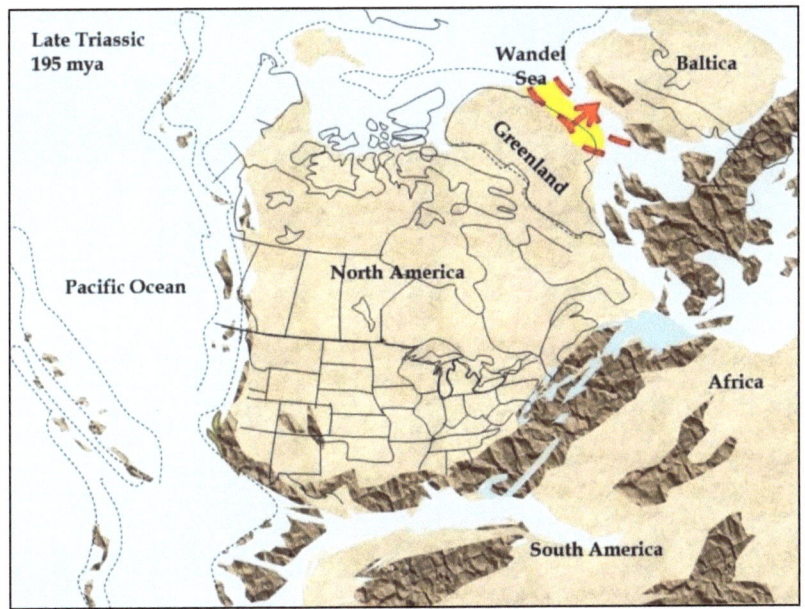

*Figure 34. Continued build out of the Western North American continental margin by island arc accretion uplifted small mountain belts in the Pacific Northwestern and Central regions. Larger mountain belts were uplifted along the southwestern and southern regions from Pangean related continental collisions.(Source: Szary, 2014.)*

Seas withdrew from Wyoming during the Early Mesozoic Era leaving behind Triassic red beds consisting of sandstone and shale belonging to the Chugwater, Red Peak, and Goose Egg Formations. Shoreline zones along a shallow coastal shelf were present. Erosion due to later uplift removed most of the Chugwater formation from the basin (**Figure 35**).

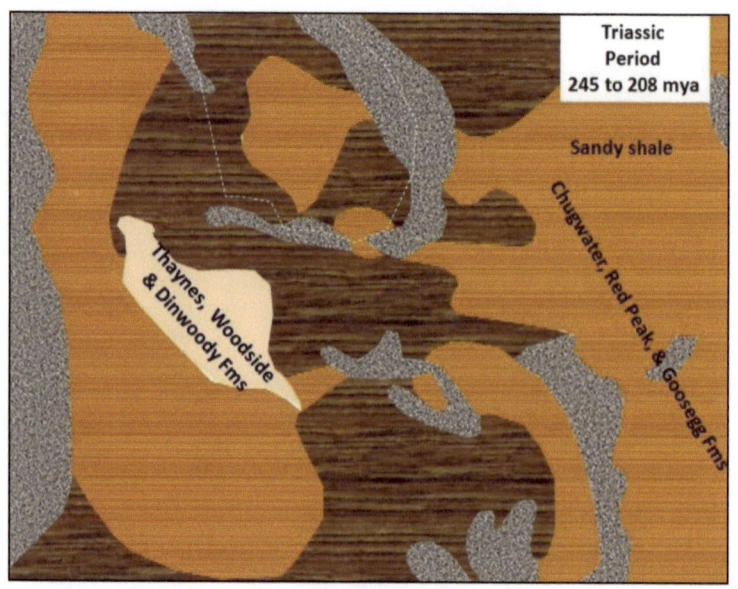

*Figure 35. Chugwater, Red Peak, and Goose Egg sandy shale covered much of eastern Wyoming. Shale covered central Wyoming. Sands formed regional belts on top of shale. (Source: Redrawn from Lageson and Spearing, 1988).*

*The Western Interior Seaway*

During the Early Jurassic Period, subduction along the western continental margin continued with island arc collisions building up mountain belts along the North American margin. Sea levels began to invade the eastern mountain belt basin when uplifting of the mountainous region to the west allowed sinking of the eastern basins to occur.

Montana, Northern Wyoming, and western North Dakota became submerged along with the western Canadian Provinces of Saskatchewan, Alberta, and British Columbia (**Figure 36**).

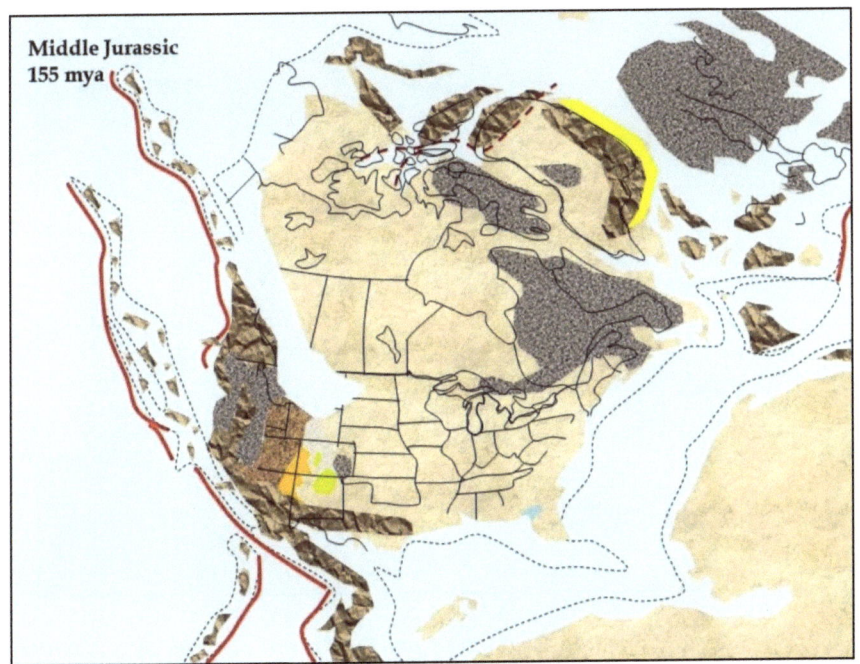

*Figure 36. Island arcs continued colliding with the western North American margin building up western mountain ranges. The Western Interior Seaway began to invade western Canada and the western states region by the rising mountains. Erosion of the central parts of the mountains, and along the eastern front shed sediments into western Wyoming. Volcanic eruptions in the four corners area covered western Colorado, eastern Utah, and northwestern Arizona. (Source: Szary, 2014).*

Jurassic floodplains covered Wyoming when rivers began to drain the landscape leaving behind mudstones, varicolored shale, sandstones, and evaporates belonging to the Gypsum Springs and Sundance Formations. The Gypsum Springs and Sundance Formations accumulated within the Western Interior Seaway (**Figure 37**). During the Upper Jurassic Period, a savannah developed, similar to present day savannah landscapes where the Morrison Formation accumulated. Perennial and intermittent streams were present at this time.

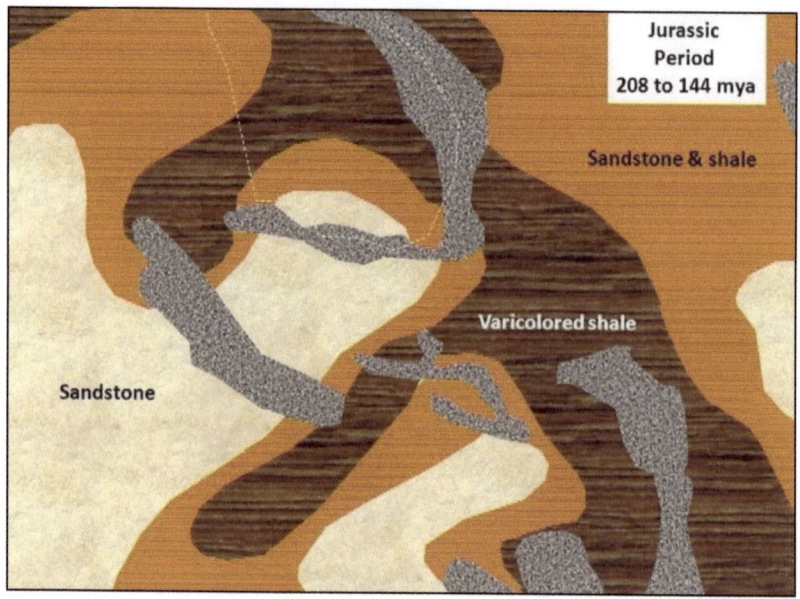

*Figure 37. Shale blanketed Wyoming from floodplains covered by sandstone and shale in the eastern and central parts, and sandstone in the western and southern parts of the state. (Source: Redrawn from Lageson and Spearing, 1988).*

During the Cretaceous Period, marine conditions returned when the Western Inland Seaway transgressed further southward from the Arctic Ocean through the western North American continent. Island arc collisions uplifted the western seaway margin, pushing the transgression cycle along the sinking eastern mountain front towards the south across Wyoming, extending southward into northwestern Colorado, 100 million years ago. Marine sediments covered the entire state.

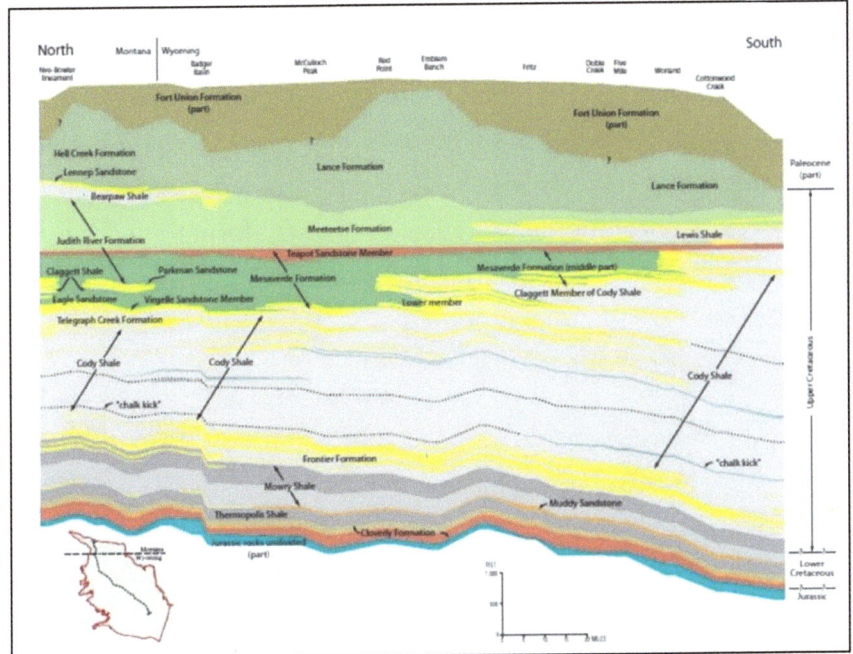

*Figure 38. General cross section showing Mesozoic and Cenozoic stratigraphic formations of the Bighorn Basin. Source: Finn, T.M., 2010.*

Sea level transgression during the Early Cretaceous Period deposited the Cloverly Formation in floodplain, fluvial, and lacustrine environments (Finn, 2010). The basal unit consists of sandstone, conglomerate sandstone, and black and gray pebble chert. Above the lower unit, fluvial and estuarine channel deposits accumulated in paleovalleys formed in a low standing basin surface which belonged to the nonmarine part of the formation at the time of initial transgression. The upper part of the formation is known as the "rusty" bed unit, accumulated in tidal flats during continued transgression.

Marine transgression continued. Thermopolis Shale consisting of dark gray to black shale interbedded with thin layers of siltstone, sandy claystone, and bentonite were deposited.

Sea levels receded partially from the Bighorn Basin. Muddy Sandstone consisting of medium to fine grained sandstone, shale, siltstone, and coal formed in fluvial, marginal marine, and estuarine environments. Thicker deposits formed within an incised valley fill complex during low sea level stand.

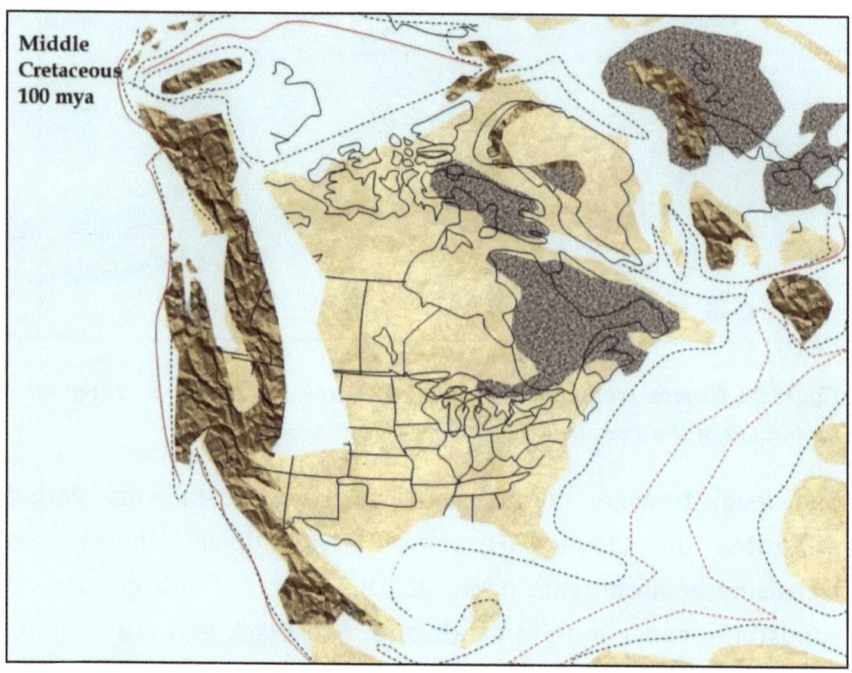

*Figure 39. The start of Laramide mountain building along the western inland sea margin continued, promoting further subsidence along the eastern mountain front. Wyoming was entirely submerged during this time. (Source: Szary, 2014).*

To the west, a narrow basin was developing in eastern Washington and western Idaho region due to the rising Laramide mountains. West of the narrow basin, the Sevier mountain building event began. The narrow basin began to flood from Pacific Ocean advancement (**Figure 39**).

About 90 million years ago, the Kula Plate subducted beneath the Oregon-California coast. Partial melting of the oceanic plate beneath the continent resulted in molten magma rising into the lower crust, forming a volcanic arc. Both Idaho and Sierra Batholiths were formed by the arc from granitic type plutonic intrusions.

The arc shifted eastward when the paleo-Pacific plate shifted from subducting east, transitioning towards the northwest direction. The shifting of the plate ended the Sevier Uplift. Erosion of the Sevier Mountains supplied sediments to the Western Interior Seaway.

By Late Cretaceous time, 85 million years ago, the Western Interior Seaway flooded the entire western region interior, submerging Montana, Wyoming, Colorado and New Mexico. The Arctic Ocean was connected to the Gulf of Mexico. Western Wisconsin was submerged when the continental craton began to subside in the northern central US. The Great Lakes region was undergoing similar subsidence while the North Atlantic Ocean began to open (**Figure 40**).

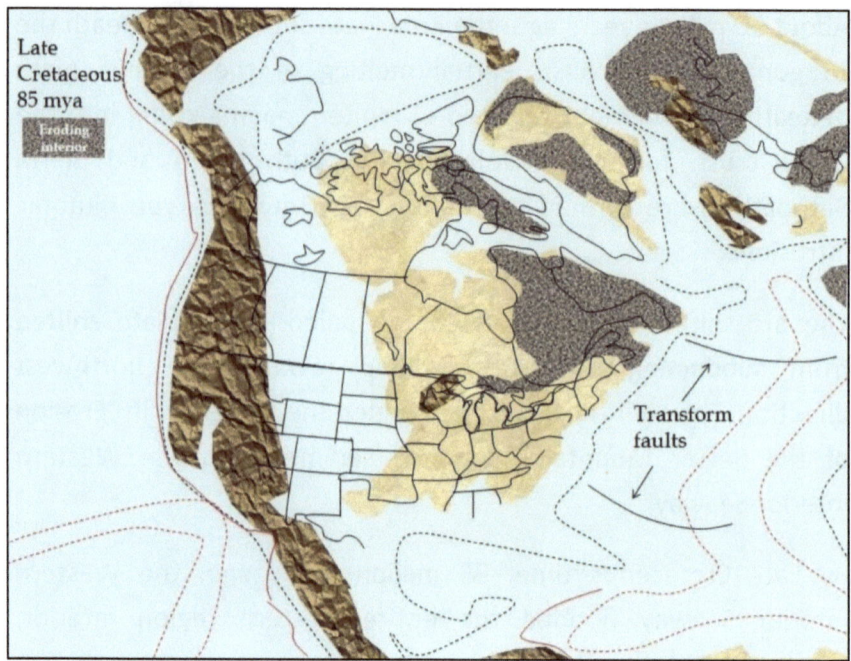

*Figure 40. The Sevier uplift forced the narrow seaway to retreat northward when the basin separating the Sevier and Laramide uplifts closed. (Source: Szary, 2014).*

Uplift and erosion began to shed sediments eastward into southwestern Montana and western Wyoming. Eastern Utah and most of New Mexico began to dry out when the interior seaway started to recede. The seaway expanded into the Canadian provinces of Ontario and northern Quebec. Erosion of the eastern Canadian provinces shed sediments towards the west into the submerged parts of Canada.

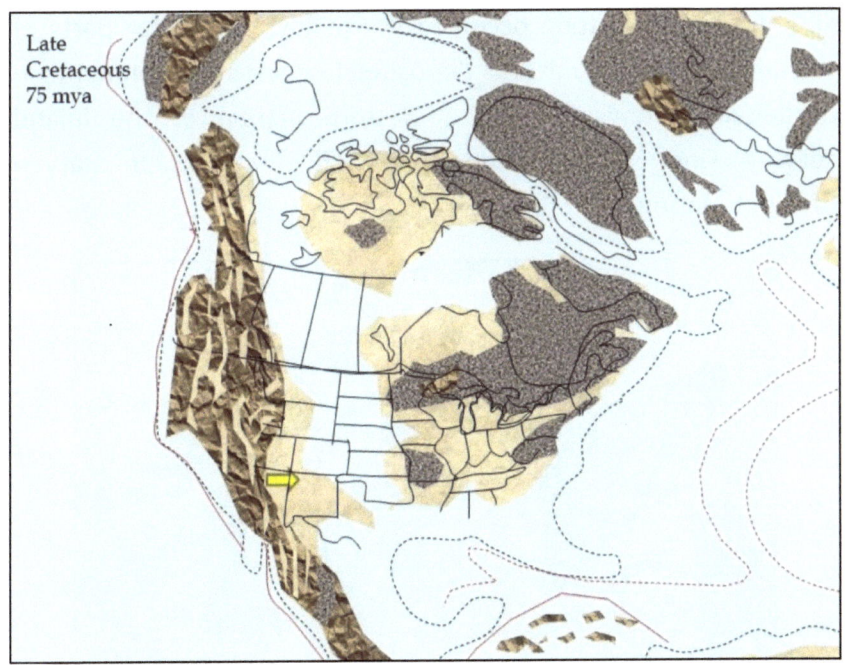

*Figure 41. Sediment shed from the eastern mountain front (yellow arrow) began to fill in the west margins of the inland seaway. Continental sediments from the eroding cratonic regions of Wisconsin and Missouri shed sediments to the west into eastern Nebraska. (Source: Szary, 2014).*

The western mountain region and central plains began to erode, 75 million years ago. Rivers drained the mountain front and central lowlands, transporting continental sediments eastward into west Wyoming, eastern Utah, New Mexico, and panhandle Texas. From the eastern lowlands, central North America shed sediments into Nebraska from Missouri and into the eastern Dakota region from Wisconsin. The Western Inland Seaway began to retreat from basin infilling (**Figure 41**).

Eastern Wyoming remained submerged while the central and western regions were covered by Mowry Formation clay rich shale in the lower part mixed with gray to tan bentonite beds. The upper part of the formation consists of siliceous shale, dark brown to black, organically rich, containing fish scales.

Fine grained sandstone occupied the middle and upper parts of the unit. The upper unit was deposited in marine environments in prodeltaic settings along the western margin of the inland seaway. Finer grained deposits formed in an oxygen starved offshore marine basin (**Figure 42**).

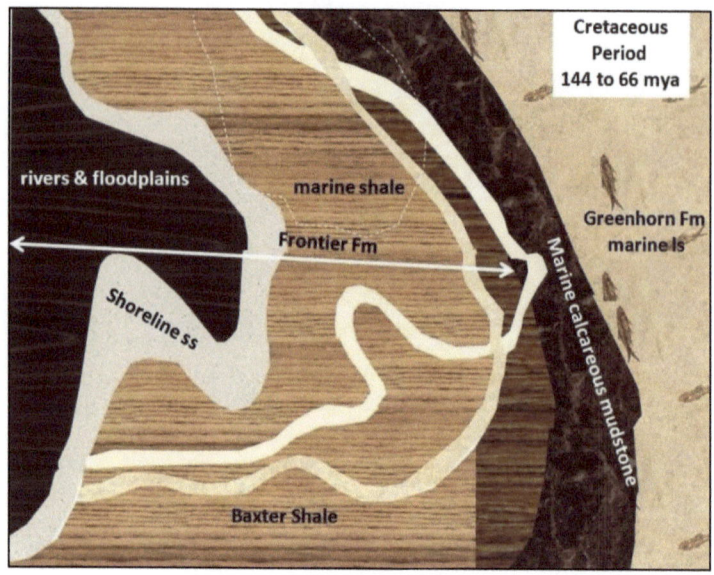

*Figure 42. Greenhorn Formation marine limestone covered eastern Wyoming. The Mowry, Thermopolis, and Cody Shale covered the central section. Frontier Sandstone stretched from the edges of the eastern marine platform to the western boundary. Marine shale acted as source rocks for oil and gas deposits. (Source: Redrawn from Lageson and Spearing, 1988).*

Frontier Formation sandstone, siltstone, shale and bentonite accumulated in a marine or marginal marine, and nonmarine setting. Sandstones belong to deltaic or shoreline origin. Nonmarine sediments consist of carbonaceous shale and coal.

Sandstones pinch out from erosion caused by marine flooding during sea level rise. A major unconformity developed in the upper part of the unit.

Marine shale belonging to the Cody Formation was deposited during a transgressive-regressive cycle occurring from lower into the upper Cretaceous Period. Sandstone interbeds pinch out against shale. Cross bedded sandstones suggest nearshore deposition occurred.

Mesaverde Formation sandstones formed in a retreating sea environment covered by nonmarine sandstone, siltstone, shale, carbonaceous shale, and coal deposited in coastal plain and marginal marine environments when the western Cretaceous inland seas retreated eastward across the Bighorn Basin. Three members are recognized: the lower member consists of very fine to medium sandstone, siltstone, shale, carbonaceous shale, and coal deposited in an eastward prograding deltaic complex. Basal sandstones grade upward into interbedded shale, carbonaceous shale, coal, and sandstone complex. The middle unit consists of interbedded sandstone, siltstone, shale, carbonaceous shale, and coal deposited in marginal marine, coastal plain, and fluvial environment along a shoreface during western shoreline retreat across the basin towards the east. Nonmarine sandstone originated as stream channel and crevasse splay deposits. The uppermost Teapot Sandstone Member light gray to white sandstone and gray mudstone developed in fluvial environments. The eastern basin contains hummocky cross bedding in the lower part suggesting a marginal marine setting was present.

The Meeteetsie Formation consists of alternating beds of sandstone, siltstone, shale, carbonaceous shale, and coal formed in a poorly drained coastal environment along the western inland sea shoreline.

The Lewis Formation consists of marine and sandy shale interbedded with sandstone grading into the Meeteetsie Formation to the west. The Lewis Formation represents the last stages of marine sedimentation prior to seaway retreat.

The Lance Formation represents the uppermost formation present in the Bighorn Basin. Interbedded sandstone, shale, and some conglomerate were deposited under fluvial conditions. The Lance Formation represents Laramide uplift where the Rocky Mountain foreland was partitioned due to thinning of the unit in the southeast and thickening near the Oregon Basin fault.

Major oil producing states follow the trend of the Western Inland Seaway from Texas northward into Alberta, eastward into Manitoba. Wyoming was part of this trend.

## *Cenozoic Sedimentation*

The Western Inland Seaway basin filled in, leaving a narrow passageway to the Gulf of Mexico. Eastern Wyoming, eastern Colorado, northeastern New Mexico, panhandle and eastern Texas remained submerged, connected to the Gulf of Mexico. Northeastern Alberta and central western Saskatchewan remained submerged in a narrow retreating seaway.

Early Cenozoic mountain building began to shed sediments from the rising uplands into adjacent basins, transported by rivers with developing floodplains and swamps.

*Figure 43. At the start of the Cenozoic Era, remnants of the Western Inland Seaway continued to occupy eastern Wyoming while the central and western regions dried out. Large scale volcanic eruptions covered parts of Montana, Idaho, Utah, and Arizona. The northern eruptions belonged to the Challis Arc in Idaho. (Source: Szary, 2014).*

Between 60 and 45 million years ago, the Kula Plate flattened out, continuing to subduct beneath the Oregon-Northern California coast. The plate extended far inland beneath Montana, Idaho, and Wyoming. A new spreading center opened up along the Pacific margin, separating the Kula Plate from the Farallon Plate. Seamounts collided with the northeastern Oregon coast. The Sierra Mountains eroded down to low hills. Further inland, central Idaho and the Absaroka Mountains formed from partial melting of the Kula Plate. The Challis Volcanic Arc formed in central Idaho (**Figure 43**).

The arc began pushing up mountains belts in Montana and Wyoming. Rising molten magma raised the Bighorn, Grand Tetons, and other Wyoming interior mountain belts during continued sinking of the Kula Plate beneath the continental margin. When the interior mountains uplifted, the interior seaway began to recede, developing river systems and floodplains, large swamps and vast lakes across Wyoming.

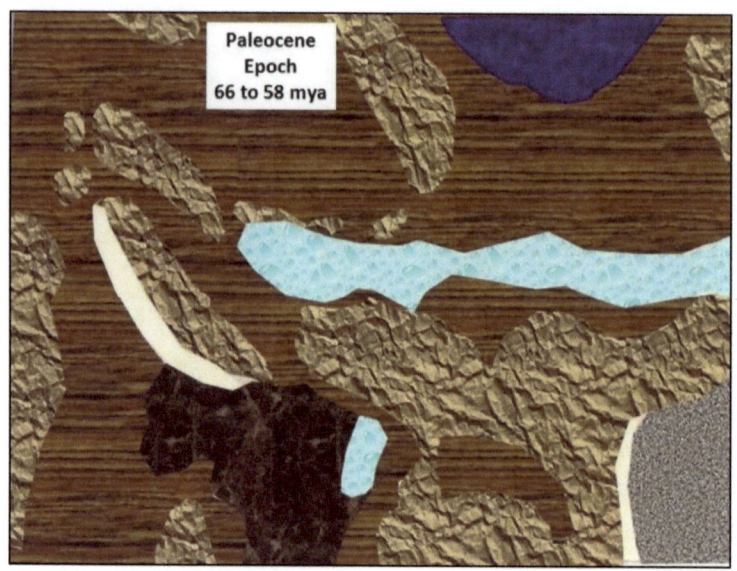

*Figure 44. The interior seaway covered the entire state with floodplain shale, regional lakes in the central part of Wyoming, and swamps in the southwestern center. The Laramide Uplift began to raise interior mountain belts which sectioned the state into basins. Volcanics extruded into southeastern Montana extending partially into the upper Powder River Basin. (Source: Redrawn from Lageson and Spearing, 1988).*

The Rocky Mountains began to uplift again in an event called the Laramide Orogeny. During Kula Plate subduction beneath the western margin, Wyoming was compressed into a series of reverse and thrust faults. Precambrian basement rock was uplifted to form many of the mountain belts known as the Wind River Range, Bighorn Mountains, and Laramie Range.

Sedimentary rocks were folded over the top and sides of the mountain belts into a foreland basin complex. Compressional forces pushed rocks into a series of imbricated thrusts and folds called the Overthrust Belt. The overthrust zone extended from central Utah into western Wyoming, western Montana through the Canadian Rockies into the Brooks Range of Alaska.

Low angle, west dipping thrust faults and folds were directly the result of subduction along the western continental margin as opposed to crustal basement rock uplift. This region later became a major oil producing province in southwestern Wyoming and eastern Utah, trapped in faulted anticlines.

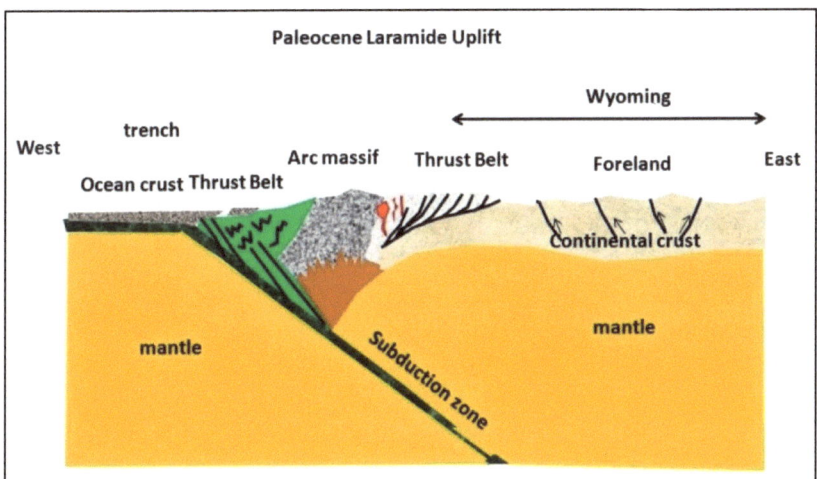

*Figure 45. Cross section of the subduction zone descent below the continental margin positioned beneath western Utah during the Laramide Uplift. The accretionary wedge thrust belt (green) developed above the trench followed by the volcanic arc massif granite (speckled). Partial melting of the upper mantle (orange) below the massif supplied magma to the arc during batholithic formation. An inland thrust system developed the Overthrust Belt of western Wyoming along with a foreland thrust region within interior Wyoming, responsible for uplifting the state mountain belts. (Source: Redrawn from Lageson and Spearing, 1988).*

Climate was wetter and humid than present day during the Paleocene Epoch. Highlands, created by the Laramide Uplift, were worn down from weathering and erosion. Thick swampy deciduous forests occupied the state, forming thick coal beds of the Fort Union Formation. Sandstone, siltstone, conglomeratic sandstone, conglomerate, carbonaceous shale and coal were deposited. Conglomerates represented alluvial fans accumulating along the northwestern margin of the basin. Braided rivers flowed northeast into the basin from the southwestern highlands. Sandstones represented river channel systems. Carbonaceous shale and shale accumulated in lacustrine and swampy environments. Continued basin subsidence occurred during the Laramide uplift. The formation is thin around the basin margins and thickest in the central part. Fossil pollen is present within the formation.

Large, shallow playa lakes covered the southwestern and central south regions of Wyoming (**Figure 44**). Wyoming became a basin and range type landscape without the typical normal faults associated with fault block topography (**Figure 45**).

Lower Eocene rocks are represented by the undivided Willwood and Tatman Formations (**Figure 46**). The Willwood Formation is composed of variegated (multi colored) shale, sandstone, and conglomerate. The Tatman formation overlies and interfingers with the Willwood, consisting of sandstone and carbonaceous shale which accumulated in lakes and swamps.

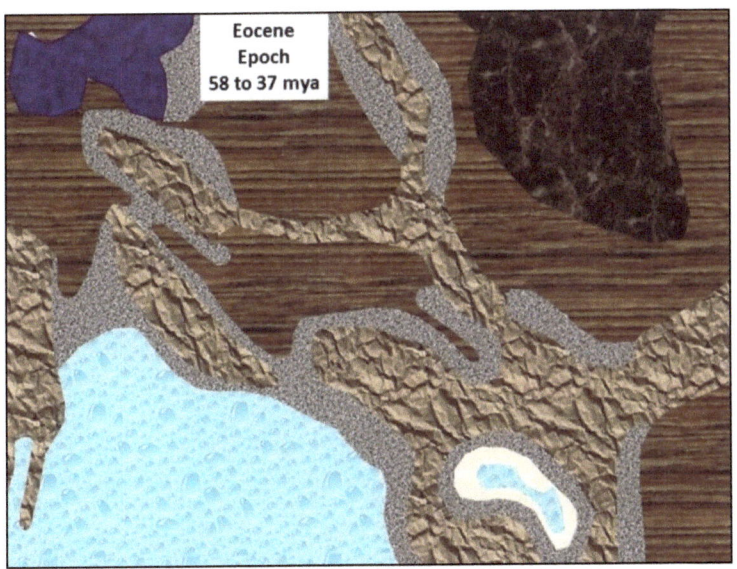

*Figure 46. The Powder River basin was occupied by a large regional swamp (brown marble) resting on top of shale left behind when the interior seaway receded during Laramide uplift. Alluvial fans were eroded and deposited along mountain belt flanks (brown stipple). The Absaroka Mountain volcanics erupted in the northwest region (purple). A regional lake occupied southwestern Wyoming and a small area in the Hannah and Laramie Basins. (Source: Redrawn from Lageson and Spearing, 1988).*

During the Oligocene, Miocene, and Pliocene Epochs, mountain belt erosion shed sedimentary deposits into the adjoining basins. Large volcanic eruptions from the Nevada and Utah Basin and Range shed ash deposits on top of Wyoming's basin sedimentary deposits. The Oligocene Wind River Formation white ash deposits are an example of the rhyolite ash deposits found in the Shirley Basin near Casper. Mountain ranges shed sediments into the adjoining basins.

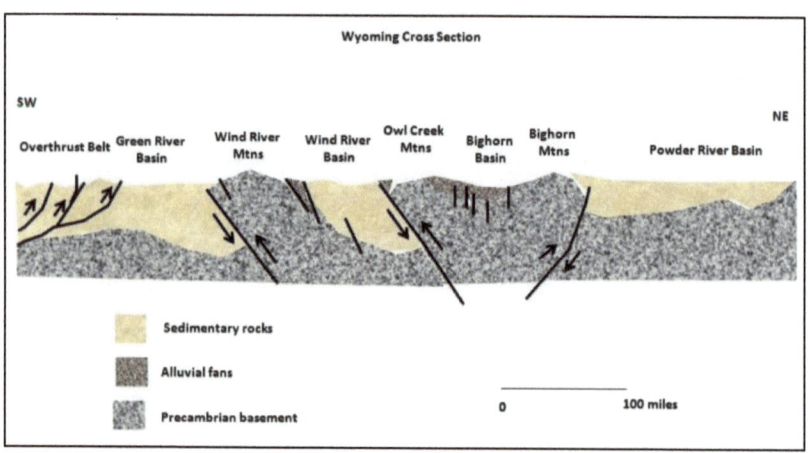

*Figure 47. Following completion of the Laramide Uplift, Wyoming became an atypical basin and range landscape with Precambrian granite mountain belts interrupted by thick sedimentary basins. (Source: Redrawn from Lageson and Spearing, 1988).*

During Miocene and Pliocene time, the Granite Mountains (Sweetwater Hills) were down faulted. Streams and rivers began to cut into buried mountain ranges, becoming incised within canyons carved into the landscape. Many rivers cut across buried mountain ranges.

During the Quaternary Period, several enormous volcanic explosions of rhyolite ash occurred when the Snake River Plain began moving eastward in Idaho. Normal faulting of the Teton Range began, accompanied by down dropping of the Jackson Hole Basin. Tensional forces began tugging the western US region apart.

Large scale uplift and arching of the intermontane region, including the Basin and Range caused rivers to cut down which accelerated erosion. High heat flow in the upper mantle was suspected of causing the regional uplift including extensional faulting and folding.

The older Laramide ranges and basins were being exposed by river erosion cutting across regional mountain ranges. For example, the Wind River is actively cutting into the Owl Creek and Bighorn Mountains instead of flowing around them.

# Chapter 3.

## *Geologic Setting of the Bighorn Basin*

The Bighorn Basin is set within north central Wyoming. The Beartooth, Absaroka, and Washakie Mountains bound the western side. The south is bounded by the Owl Creek and Bridger Mountains. The Bighorn and Pryor Mountains bound the eastern basin. The northern basin is connected to the Crazy Mountain Basin of south central Montana. The basin center is characterized by low, flat, and dry plains. Mountains are forested. Summits are snow-capped rising to greater than 13,000 feet. The Bighorn, Shoshoni, Greybull, and Clarks Fork Rivers drain mountains carrying sediments into the basin. Rivers discharge into the Yellowstone River in Montana. The northern Bighorn Mountains were cut into, forming a narrow canyon by the Bighorn River.

Folds and faults are exposed in the basin, originating from the Laramide Uplift. Geologic structures can be recognized by three geologic zones which characterize the basin (**Figure 48**). Zone 1 belongs to the high mountain rim surrounding the basin. The rim consists of uplifted Paleozoic sedimentary rock (blue) where portions of the rock were eroded to expose Precambrian granitic and plutonic rocks (pink shades). Mountains were uplifted by thrust faults during the Laramide Orogeny, 60 to 55 million years ago. Precambrian basement rocks were thrust along the eastern margin along the Beartooth Mountains in the northwest corner. Younger rocks were exposed beneath the Eocene Absaroka volcanics. Along the eastern boundary, the Bighorn Mountains were thrust westward over the basin margin at the north and south ends of the range.

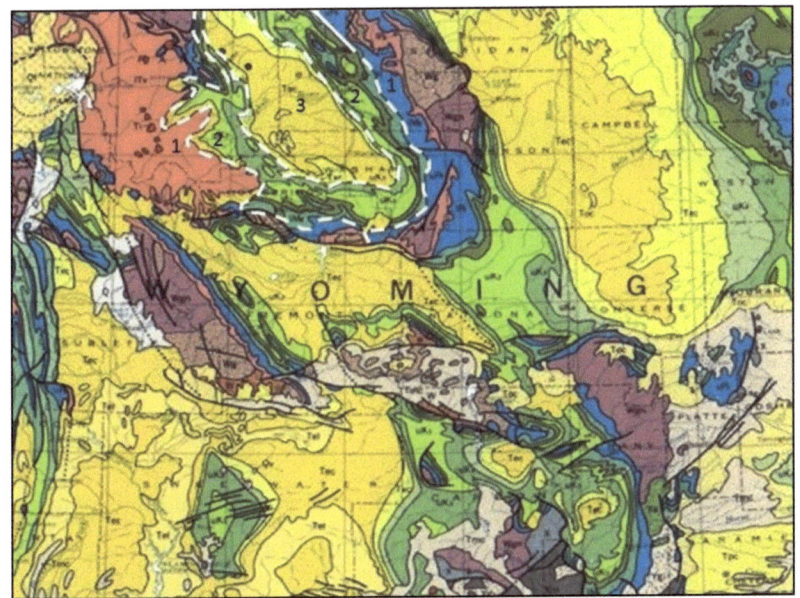

*Figure 48. Geologic map of Wyoming showing the Bighorn Basin outlined by dashed lines (white). Numbers represent the various zones delineated by the Wyoming Geological Survey. Source: Wyoming Geological Survey.*

Zone 2 is a platform composed of Mesozoic aged sedimentary rocks consisting of anticlines and synclines encircling the basin. The platform forms a bench between the high mountain ranges and flat basin. During Laramide uplift, mountains rose and the basin subsided. Displacement developed a zone of faults which uplifted the platform. Many anticlines were formed asymmetrically, trapping oil and gas deposits. Steeper limbs face outward towards the mountains. Reverse and/or thrust faults bound the platform and are buried deep, providing the mechanism for raising Precambrian basement rock. When the basement uplifted, overlying sedimentary rocks were folded. Anticlines are recognized by the bright red Chugwater formation, exposed along the fold core and flanks.

Anticlines act as traps for Bighorn Basin petroleum. Oil was generated from compaction and heating of carbon rich, marine Paleozoic sediments belonging to Phosphoria Formation black shale, for instance. The process began millions of years prior to the Laramide uplift. Theories suggest oil migrated upslope from source areas in western Wyoming and eastern Utah into the Bighorn Basin. Either during or after Laramide uplift, oil migrated by secondary movement into anticlines occupied by sandy formations such as the Pennsylvanian aged Tensleep Sandstone.

Zone 3 belongs to the central basin trough. Younger Cenozoic aged rocks make up this zone. The western edge of the basin was compressed eastward over the central basin trough along the Oregon Basin Fault. The fault dips 30 degrees westward over the central trough. The Paleocene Fort Union and Eocene Willwood formations outcrop in the central portion of the basin. The Wasatch Formation (darker yellow) overlies both formations except for patches located in the southwestern corner.

### The Petroleum Trap

There are basically two types of petroleum traps encountered in the Bighorn Basin: structural and stratigraphic. Traps occur in Permian and Cretaceous source rocks and reservoir systems. Structural traps include folds and faults which act to seal migrating petroleum from moving further upslope within permeable formations due to impermeable formations. The following table provides a summary of petroleum source rocks and trap types within the Bighorn Basin.

| Formation | Trap Type | Age |
|---|---|---|
| Phosphoria Formation | stratigraphic | Permian |
| Tensleep Formation | Paleo-topographic | Pennsylvanian |
| Greybull-Cloverly-Muddy SS | Stratigraphic | Lower Cretaceous |
| Bighorn-Darby Wedge | Stratigraphic Pinch out | Ordovician-Devonian |
| Flathead-Lander SS | Stratigraphic | Cambrian |
| Madison LS | Stratigraphic | Mississippian |
| Darwin-Amsden SS | Stratigraphic | Permian |
| Triassic-Jurassic | Stratigraphic | Mesozoic |
| Basin Margin | structural | Cretaceous |
| Deep Basin | structural | Cretaceous and younger |

## *The Basin Margin Anticlinal Trap*

Petroleum traps are set within anticlinal and dome type structures formed during the Laramide Uplift. Many folds are faulted along the center and at plunging noses within the shallow margins of the Bighorn Basin. The top of the Tensleep Formation marks the boundary in the central and eastern parts of Wyoming. The western boundary is marked by the eastern edge of the Eocene Absaroka Volcanics. Reservoir rocks range in age between Cambrian and Cretaceous, consisting of Flathead, Bighorn, Jefferson, Madison, Amsden, Tensleep, Phosphoria, Ervay, Dunwoody, Crow Mountain, Chugwater, Cloverly, Dakota, Greybull, Lakota, Muddy, Frontier, and Mesaverde Formations. Production occurred from the Madison, Tensleep, Phosphoria, and Frontier Formations. Sandstone is the primary reservoir type rock for each of these formations. Madison, Jefferson, and Phosphoria Formation carbonate rocks also produced petroleum.

Petroleum was locally formed without distant migration during the Laramide Uplift. It is possible petroleum was generated prior to Laramide influences in Western Wyoming where migration occurred during Laramide uplift.

Cretaceous source rocks probably reached maturity by Paleocene time in the deeper parts of the basin followed by burial of younger sedimentary sequences. Structural traps formed during Laramide uplift which resulted in final petroleum reservoir entrapment.

Trapping mechanisms occurred when both anticlines and domes were faulted along the central fold and along plunging noses during Laramide uplift. Best structures were formed around the shallow basin margins between a few hundred to 12,000 feet deep. Within folded structures, impermeable beds sealed petroleum where Paleozoic and Mesozoic reservoirs were separated.

Petroleum reservoirs must have at least four essential elements in order to trap oil, gas, and water (Levorsen, 1967). The reservoir rock is the location where petroleum resides. The reservoir rock must have pore spaces which interconnect. Pore spaces are where the petroleum is held and stored in the reservoir rock. The third element is the petroleum itself, occurring in the form of oil, gas, and water either in motion or capable of moving through interconnected pore spaces. The fourth element is the trap which is the place where petroleum is prevented from further movement.

Oil and gas are lighter than water. This allows the petroleum to move through water in both vertical and horizontal directions until it is trapped by an impervious rock or rock with reduced permeability. The impervious rock unit which overlies the reservoir rock is called the roof rock, or cap rock. When the roof rock is concave, which prevents oil and gas from escaping while allowing petroleum to pool, the barrier is called a structural trap. When permeability is reduced in a horizontal direction due to changes in rock facies, truncations, or other stratigraphic related changes in combination with the roof rock, a stratigraphic trap is formed. The simplest kind of structural trap is called an anticlinal trap.

Three types of traps are recognized: 1) structural traps; 2) stratigraphic traps; 3) combination traps. Structural traps are those whose upper boundary is concave when viewed from below formed by local deformation including faulting, folding or some combination of the reservoir rock. The edges of the pool are defined by the intersection of the underlying water table with the roof rock overlying the deformed reservoir rock.

Stratigraphic traps are defined by some variation in stratigraphy or lithology of the reservoir rock. Facies changes, variation in porosity and/or permeability, or up-dip structural termination of the reservoir rock, are the most commonly formed stratigraphic traps.

The pool may rest above an underlying water table or may completely fill voids in the reservoir rock with the absence of producible reservoir water. The most important element is the flow of water down dip through restricted permeable rock which forms a stratigraphic barrier to the up dip movement of petroleum.

Combination traps are defined by both structure and stratigraphy. The position of a pool within a trap depends on formation water movements. Where formation water movement is absent, the pool is positioned at its highest point in the trap. Where formation water movement occurs, the pool migrates for varying distances down the side of the trap.

### Structural Traps

Folding and faulting are the primary causes for trapping petroleum. Anticlinal traps extend vertically into the subsurface through a considerable thickness of sedimentary formations. Traps will form in all potential reservoir rocks affected by the geologic structure. Nearly all petroleum traps are composed of deformed reservoir rock. Folds may be the only source of the trap but other factors may also play a role. Structural combination with stratigraphic and fluid conditions may also influence petroleum trapping mechanisms.

Geologic structure at the surface may be quite different in the subsurface. Folds may shift laterally or may lose closure with depth becoming an ineffective trap.

The pool may also become displaced in the down gradient direction of the fold by moving water within the reservoir rock. In the subsurface, folding, faulting, and other deformation occurs similar to that observed at grade. Structural features may exhibit combinations in many ways. Structural classification includes folding, faulting (normal and reverse), fracturing, salt plug intrusions, and any combination of the proceeding structures.

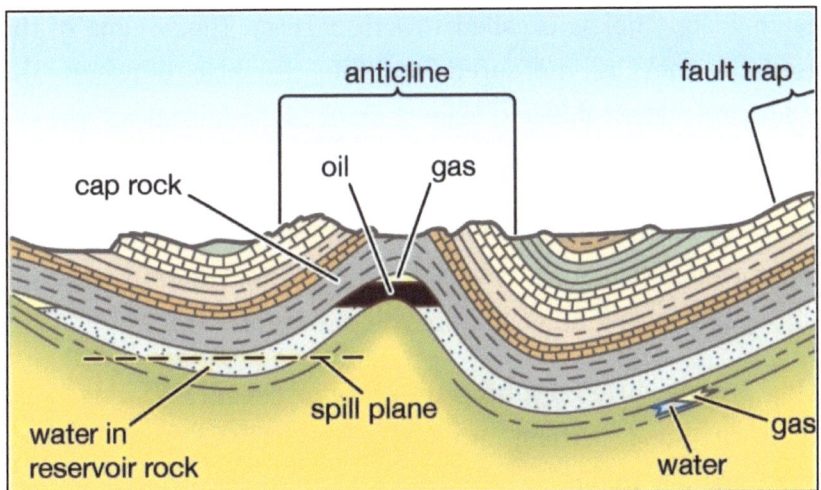

*Figure 49. Anticlinal folds are the most common traps where petroleum accumulates. Reservoir rocks provide enough permeability to allow petroleum to migrate to the highest point within the fold structure. Impermeable rock, above the reservoir rock traps petroleum within the fold. Source: USGS.*

*Folded Traps*

Folds consist of widely varied shapes including low profile circular shaped domes, long narrow anticlines which may be symmetrically or asymmetrically shaped, or overturned. The vertical distance between the highest part of the fold to the lowest closed contour is called *structural closure*. Structural closure of traps are referred to as a datum referenced to feet above or below mean sea level.

Capacity of a folded trap to hold petroleum depends on structural closure, thickness of the reservoir rock, effective porosity, reservoir pressure, and fluid flow conditions through the reservoir rock (**Figure 49**).

Structural closure may range from a few feet up to thousands of feet. The height to which a fold rises above the regional slope measured from the highest point of the fold to the projected regional slope below is called *structural relief*. The volume of the reservoir rock is measured by the distance between the underlying water table and highest point of the fold structure.

Folds are present in virtually every trap regardless of whether the trap is structural or stratigraphic. Where folds are the principle trap forming structure, it is classed as a folded trap although faulting and stratigraphic features help to form closure.

Folds are produced by horizontal compression, shearing, drag folds, dips and settling around buried hills, diapiric folding, superimposed folding, and domes resulting from intrusive salt plugs. Folding may occur in a single event or through multiple events. Faulting usually accompanies folding but is difficult to detect in the subsurface.

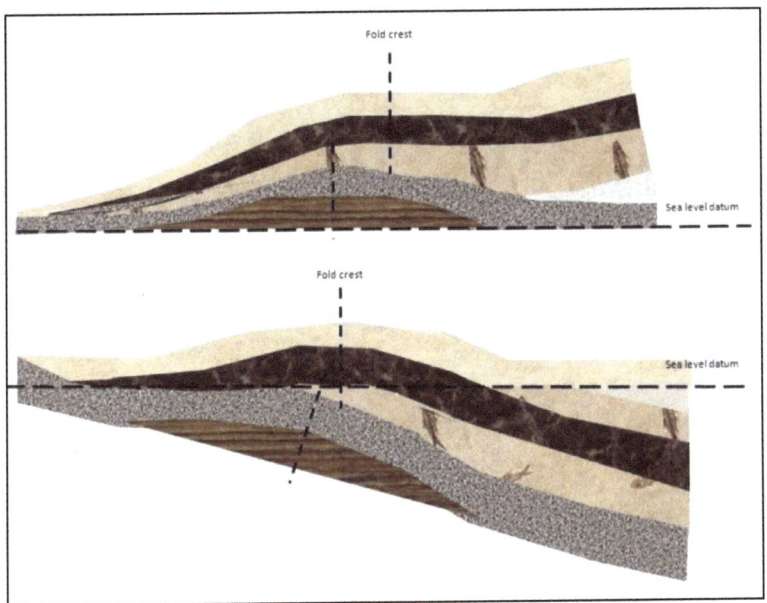

*Figure 50. Structural closure gains and losses shift with depth when strata converge. The crest of the fold changes between the upper and lower fold. Drawing modified from Levorsen, 1967.*

Fold structures change with depth in size, shape, amplitude, or lateral shifting between the surface, and vertically between shallow depths to the reservoir rock. Folding at grade or at shallow depths is not a reliable guide for estimating depths to petroleum pools because fold crests do not remain consistently parallel to the land surface for deeper fold structures. Levorsen (1967) describes structural fold types in more detail, summarized in the following section.

## Fold Pattern Shifting

*Convergence of Intervening Strata.* When strata between the surface or shallow depth and the reservoir rock converge regionally, deep folding within the reservoir rock may shift position with depth in the direction of convergence. Structural closure either increases or decreases with respect to sea level (**Figure 50**).

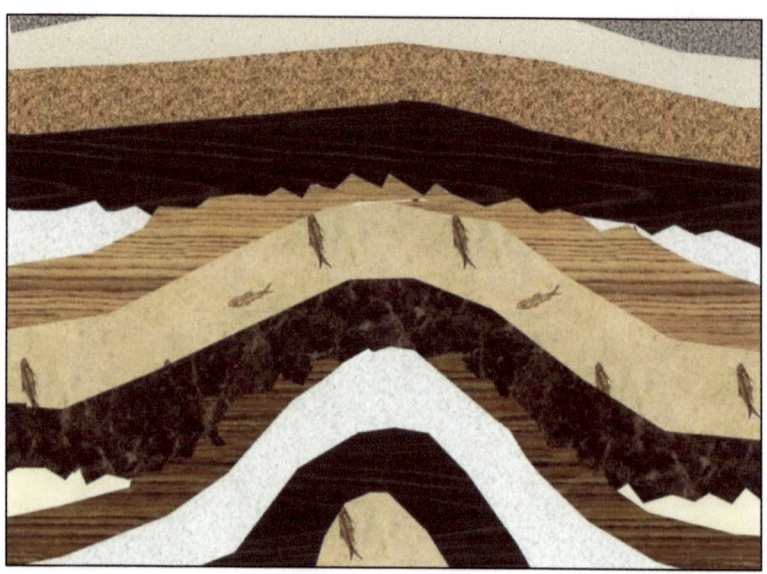

*Figure 51. Three episodes of folding are represented by the diagram. Folding occurred before the lower unconformity surface formed at the dark brown marble layer. A second folding event occurred later before the upper unconformity formed below the cork textured strata at the surface. A third folding event occurred with less intensity after the cork layer was deposited. Drawing modified from Levorsen, 1967.*

*Repeat Episodes of Folding.* Multiple episodes of folding produce greater structural relief with depth. Formations become thinner on the crest of the fold than on the flanks. Thinning may persist or become intermittent throughout a portion of the geologic column. Folding is more intense below unconformities (**Figure 51**).

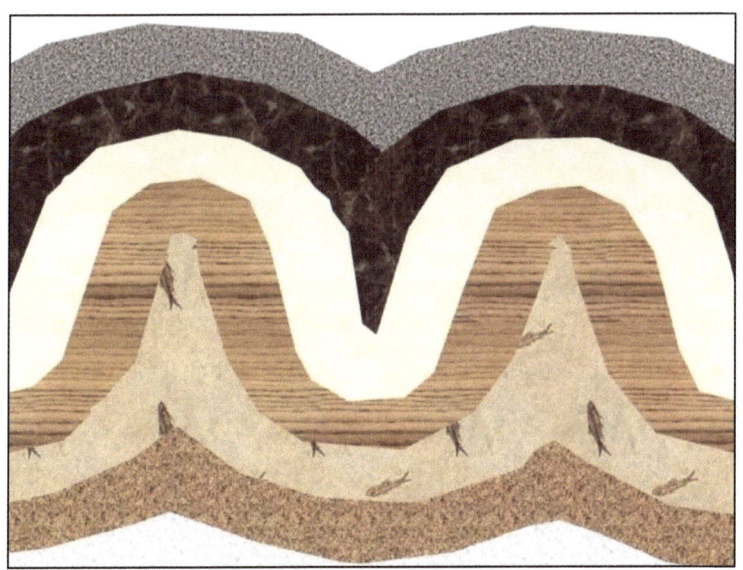

*Figure 52. Parallel folds increase structural closure with depth where reservoir rocks occur several miles below grade. With sufficient depth, anticlines resulting from parallel folding eventually disappear. Drawing modified from Levorsen, 1967.*

*Parallel Folding.* Shale formations folded into low relief anticlines and domes are expected to fold normally. Normal folding occurs when the fold in the reservoir rock is parallel to the surface and to folds in the intervening beds. Bed thicknesses do not change during the folding process but folding becomes more acute with depth. Where reservoir rocks are two or three miles below grade, folding is likely to occur sharper than at shallower depths (**Figure 52**).

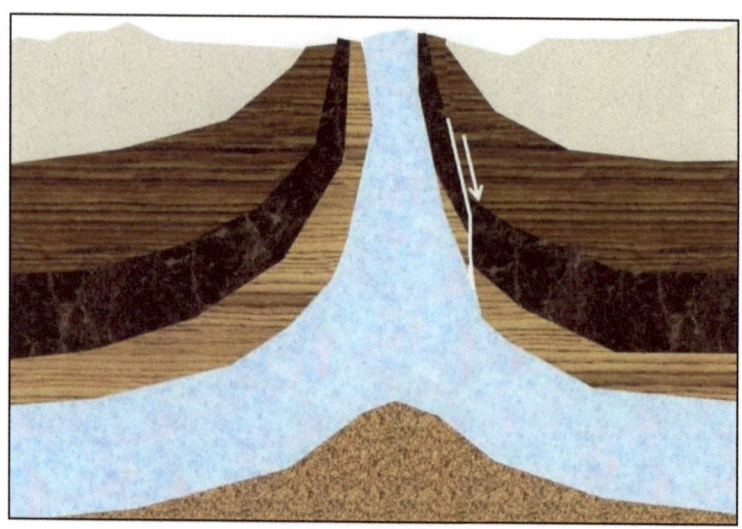

*Figure 53. Salt intrusions are squeezed upwards by vertical compression exerted on the deposit by heavier shale and muds in the subsurface. Drawing modified from Levorsen, 1967.*

*Discordant or Diapiric Folding.* Incompetent beds, or weak strata may underlie or overlie competent beds. When a competent reservoir rock is overlain by great thicknesses of soft, incompetent rock a great difference may emerge between folds occurring at the surface or at shallow depths than folding occurring in the reservoir rock (**Figure 53**).

When incompetent formations underlie competent formations, a central core of older competent material is squeezed or injected upwards through the crest of an anticline. These types of folds are called diapirs or piercement folds. Petroleum is generally trapped in the gently dipping formations adjacent to the steeply dipping axial rocks.

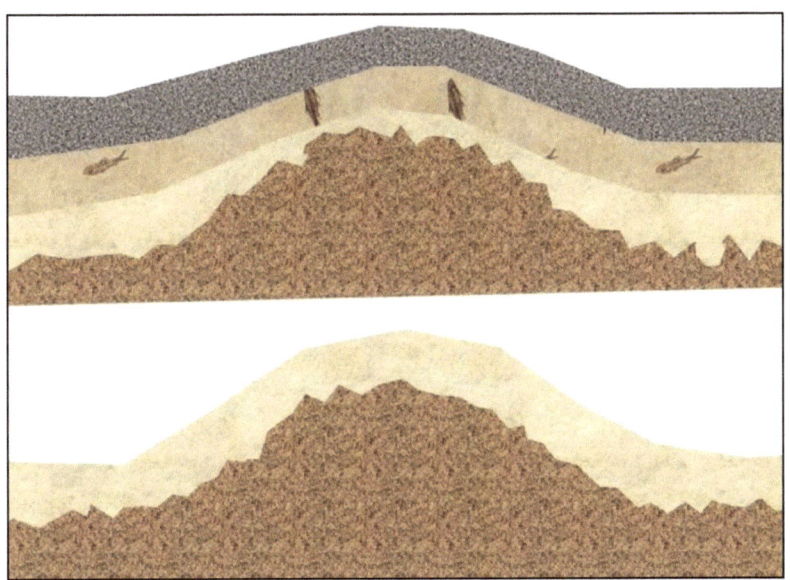

*Figure 54. Buried hills (upper graphic) show thinning of layers occurring above the unconformity with topographic relief. Where buried hills are absent, uniform thicknesses of layers occurring above the unconformity occur with no topographic relief. Drawing modified from Levorsen, 1967.*

*Buried Hills.* Occasionally, some folds arch over buried hills. Buried hills are not actually hills or topographic highs when overlapping sediments are deposited. They are folded surfaces exposed to erosion which form unconformities. Many are buried anticlines which were later eroded with little to no topographic relief. They become buried when younger formations are deposited on top of the eroded surface, later to be refolded along with the overlying formations (**Figure 54**).

A true buried hill consists of a topographic high, a bioherm, organic reef, or a resistant lens composed of sands, gravels, clays, or shale. These types of folds may form under two types of circumstances. Compaction around the edges of the hill may be more extreme than over the hill crest. The overlying sediments would drape over the edges of the hill forming a dome or anticline.

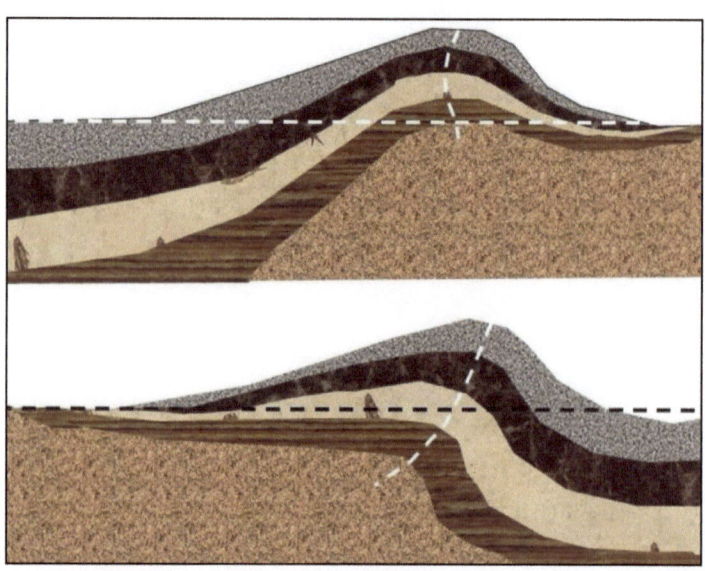

*Figure 55. The upper diagram shows formation thinning in the direction of the steep side of the fold. The crest of the fold shifts in the direction of formation thinning. The lower diagram shows beds converging toward the lower dipping direction of the asymmetrical fold. The fold crest shifts in the direction of thinning. Drawing modified from Levorsen, 1967.*

*Asymmetric Folding.* Asymmetrical folds cause the fold crest to shift towards the direction of the fold flank with the lowest dip. Where the depth to reservoir rock is deep (two to three miles), shifting may be considerable when there is a large difference between dip of the opposing flanks (**Figure 55**).

*Pre-unconformable Deformation.* Folds and faults occurring below buried unconformities often are not exposed at the surface due to burial by thick deposits which accumulated after erosion has occurred. Traps become buried below the older surface of erosion (**Figure 56**).

*Figure 56. Pre-unconformable traps become obscured by post unconformable formations. The lower part of the section represents stratigraphic deposition occurring before folding. Faulting separated the petroleum trap before the upper surface was eroded. The upper formation was deposited and later titled, burying the lower structural trap. Drawing modified from Levorsen, 1967.*

*Overriding Thrusts.* Thrust faults tend to obscure underlying structures, trapping several pools in structures concealed by overriding sediments.

*Displaced Pools.* Pools may be displaced along varying distances downward alongside the original trap. Sometimes, the original high point in the reservoir rock will retain petroleum but may become barren when displacement occurs. Displacement is caused by fluid potential gradients which allow water to migrate through reservoir rock.

Normal, reverse, and thrust faults in the reservoir rock may wholly or partially trap petroleum. Many structural traps are faulted but faults may not be the primary trapping mechanism. Combinations of structure and faulting are more common where folding, tilting, and arching of strata vary with stratigraphy and permeability differences (**Figure 57**).

Petroleum seepages are often associated with faults which extend back to grade. Faults may also act to connect subsurface reservoir rocks with the surface. Seepages suggest the ground water potentiometric surface is greater than the land surface elevation.

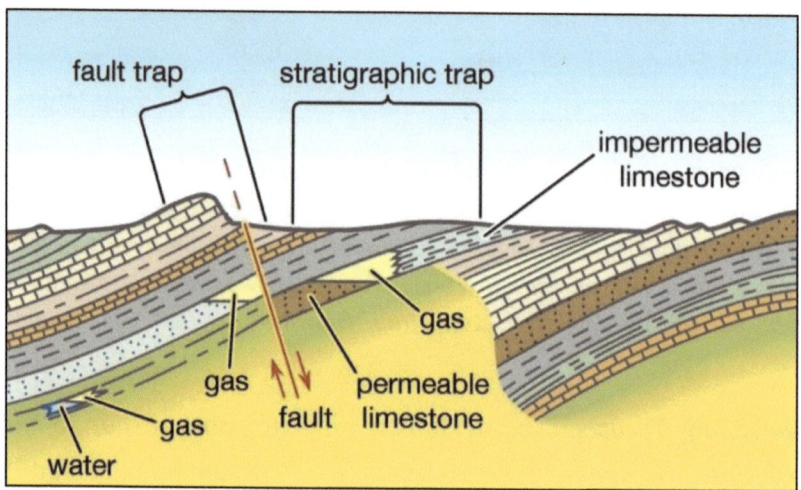

*Figure 57. Normal, reverse, and thrust faults trap petroleum by closing off reservoir rocks through seals occurring along the fault trace due to the presence of impermeable gouge material. Where gouge material is absent, or coarse grained, petroleum may seep along the fault eventually reaching the surface.*

Lack of seepage suggests the ground water potentiometric surface is below the land surface elevation. Where faults form a boundary plane of pooling petroleum, higher fluid potential within the fault and up dip across the fault creates a hydraulic barrier to petroleum movement. Faulting and hydrodynamic conditions forms the trap which holds the pool.

*Normal Faults.* Normal faults combined with a regional monocline dip forms a trap. There may be a single curved fault, intersecting faults, or a combination of many faults. Normal faulting combined with low folding forms many pools. When folding becomes more acute, the trap becomes more clearly defined, usually located on elongated anticlines and domes.

Faulting breaks pools up into separate traps. The fault plane becomes the mechanism which prevents and seals petroleum from migrating further up dip through the reservoir rock or through the fault plane itself.

Minor faults occasionally follow fractures coupled with subsurface effects related to folding. When erosion and overburden removal occurs, stresses are brought closer to the surface. Stresses are relieved by faulting.

Normally faulted petroleum pools accumulate on the upper side of the fault. Pools tend to escape when formed on the lower side due to migration around the up dip ends of the fault plane.

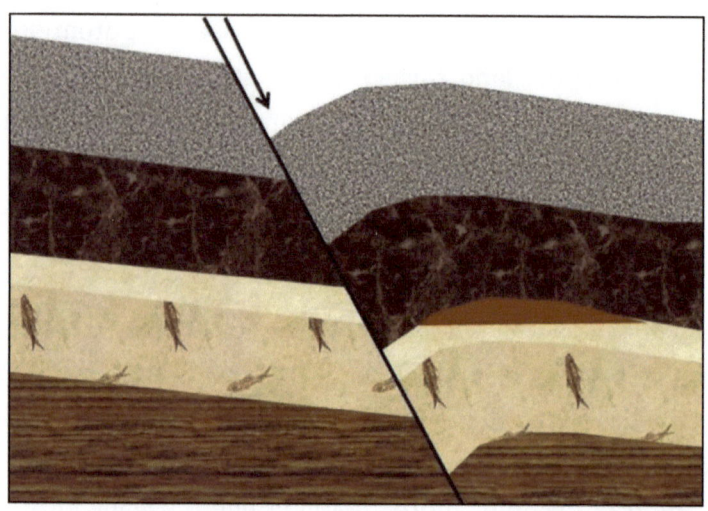

*Figure 58. Petroleum pool accumulation on the down dropped side of the fault plane, characteristic of traps located along the Gulf Coast of Texas and Louisiana. Petroleum is represented by medium brown color. Drawing modified from Levorsen, 1967.*

In the case along the Gulf Coast of Texas and Louisiana, pools are generally located on the down dropped side of the fault plane. Normal faults dip to the south, seaward from the shoreline. Traps are formed by anticlines exhibiting structural closure which occur coincident with fault planes and parallel to the down dropped side of the faults. Faulting plays a very minimizing role in formulating the trap itself. Faulting is genetically related to folding (**Figure 58**).

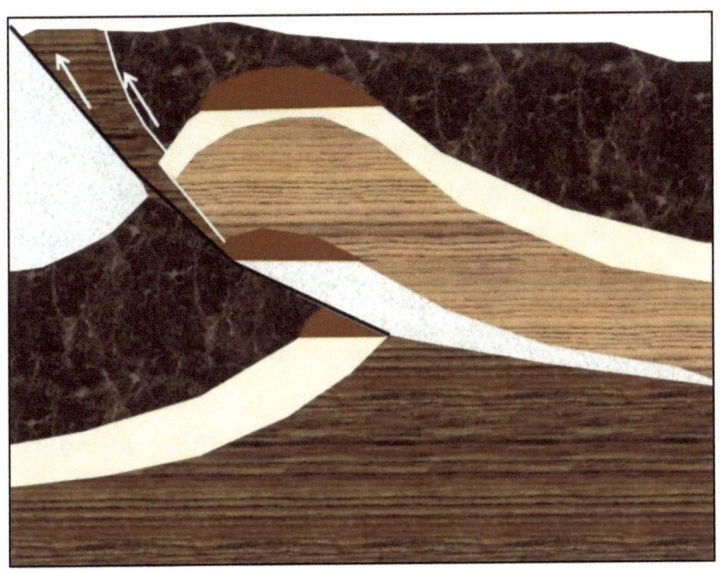

*Figure 59. Generalized thrust fault diagram showing petroleum traps (medium brown) associated with offsetting reservoir rocks. Petroleum is represented by medium brown color. Drawing modified from Levorsen, 1967.*

*Reverse and Thrust Faults.* Traps formed under reverse or thrust faulted regions form either above or below the fault plane. Traps may become bounded by the fault plane itself but generally form by folding associated with faulting (**Figure 59**).

*Fracture Traps.* Fractures within reservoir rock result in increasing porosity and permeability. Flat lying rocks may contain petroleum traps formed by fracturing. Where fractures diminish and eventually play out, petroleum trapping no longer exists.

Structural traps generally produce petroleum containing a free gas cap positioned on top of the water oil contact. Reservoir energy is driven by water pressure exerted on the reservoir by depth. Structural traps tend to extend vertically through the structure with thick sections of potentially productive rocks. Large structural closures are required for the trap to be effective.

A stratigraphic trap is defined by changes in lateral variability within the reservoir rock lithology. Permeable reservoir rocks may change into less permeable rock or into an impermeable rock horizontally. Unconformities and overlapping of reservoir rocks by younger formation units also form traps. Many traps consist of structural and stratigraphic combinations (**Figure 57**).

Two classes of stratigraphic traps are recognized: primary and secondary. Primary stratigraphic traps are those formed during deposition or diagenesis of the rock including lenses, facies changes, shoestring sands, and reefs. Secondary stratigraphic traps are those which formed from later causes including solution, cementation, and unconformities.

*Primary Stratigraphic Traps*. Primary traps form within a specific depositional environment which influences the reservoir rock properties and the conditions under which the rock forms. Effective pore space and impervious upper bounding surface of the trap result from primary sedimentation processes.

*Lense type traps* consist of porous and permeable clastic rock surrounded by impermeable sediments. For example, a river channel deposit surrounded by clayey and silty floodplain deposit (**Figure 60**).

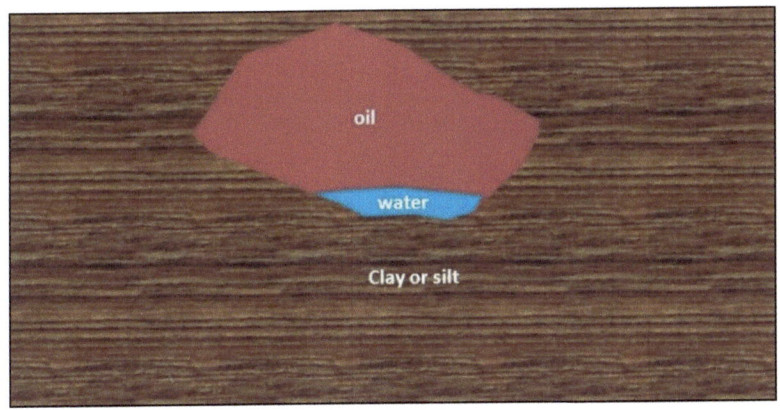

*Figure 60. Typical lens shaped trap surrounded by impermeable rocks. Drawing modified from Levorsen, 1967.*

Sandstones, arkose, coquina, and more rarely weathered, brecciated, and redeposited basalts and serpentine derived sediments may form lenticular shaped petroleum traps. Turbidites compose many of the lenticular shaped deposits when deposited on the seaward side of rapidly filling depositional basins. Boundaries may either be sharp or gradational.

*Figure 61. Changes in permeability between a water filled permeable sand contacting an impermeable clay or silt forms petroleum traps, called lithofacies traps. Drawing modified from Levorsen, 1967.*

Facies changes are lateral gradations within a formation or group of rocks resulting from deposition of rocks occurring at the same time. Where differences are caused by lithology, a lithofacies change occurs. Where fossil contents change, we have a biofacies change. Many traps result from changes in lithofacies resulting in permeable to impermeable rocks. When sandstones grade into shale at the upper monocline edge, or permeable dolomite yields to impermeable limestone, the up dip edge of the permeable formation may mark the petroleum trap (**Figure 61**).

Petroleum may completely fill the porous part of a sand lens at the high portion of the lens. If one primary trap is discovered, the phenomenon may be regional, containing many traps similar to the discovery.

Sand patches, lenses, sandy zones, bars, and channels may occur in complete randomness. These types of reservoirs may become truncated by other formations containing similar lithologic types.

Strandline pools are associated with shoreline environments where regional facies change from permeable to impermeable rocks. Serial or grouping of pools may occur along long linear trends.

*Shoestring sand traps* are long, narrow sand deposits consisting of permeable sands surrounded by impervious clays and shale. Some shoestring deposits may be of channel origin, and formed by offshore sand bars.

Offshore sand bars are characterized by flat bases with an upper concave surface when viewed from below. The sides of the deposit are relatively straight. The seaward side is in sharp contact between sand and shale. The lagoon side grades into shale and clay, has a muddy texture, and is impermeable.

Grain size sorting, composition, and texture run parallel to the edge of the deposit and are uniform in texture. Whereas in channel filled deposits, sorting is more variable. Petroleum production is more uniform in offshore deposits than channel filled deposits.

*Channel fillings* with sand, gravel, and clastics form when streams meander back and forth across a floodplain leaving remnants of the old channel when the channel becomes clogged by sand and gravel. At entrances to the ocean, distributary streams balance water discharge, sand, gravel, and silt. Tidal flats, deltas, and deltaic deposits contain sand and gravel filled channels which are bypassed by later stream loads which discharge along the shoreline. Channel fill may combine with other channel deposits to form widespread deposits.

Channel deposits are recognizable in the subsurface by a convex base deposit when viewed from below due to sands filling in a pre-existing valley. Sediments within the channel cover a wide range of grain sizes, composition, and texture. Deposits also form a sinuous, meandering pattern occasionally with curved outlines similar to river oxbow lakes and channels. Fossil channels cut into underlying rocks which later fill in with sand, conglomerate, and other clastic deposits.

*Carbonate Rock Facies*

The most common type of carbonate rock consists of dolomitized limestone. Magnesium carbonate replaces calcium carbonate which reduces the volume of rock resulting in an increase of porosity and permeability. Petroleum traps appear in sandy or cherty facies encased within carbonate rock. Other traps form in recrystallized clastic lenses composed of shell, coquina, oolites, or carbonate fragments.

Organic layers of these types of limestone are called biostromes formed in place by organisms attached to the carbonate reef structure or washed in by currents and waves.

*Organic Reefs*. Traps formed in organic reefs consist entirely of organic material which grew in place or became complex mixtures of interbedded original and detrital organic material and debris. Bioherms are dome like, mound like, or circumscribed masses built exclusively by corals, stromatoporoids, algae, brachiopods, mollusks, and crinoids enclosed within rock with different lithologic types.

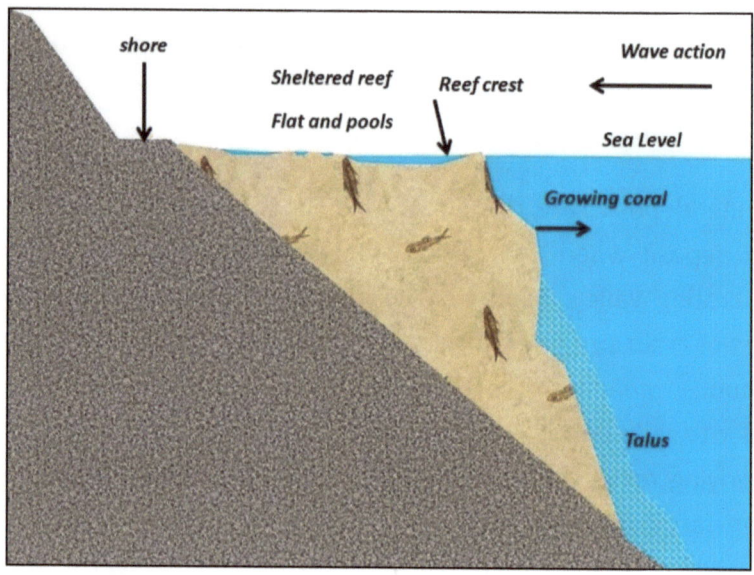

*Figure 62. Anatomy of a fringing reef with typical parts making up the reef complex. Drawing modified from Levorsen, 1967.*

A typical fringing reef is attached to the shoreline, exposed during low tidal cycles in the form of a shelf or flat a few feet to a quarter of a mile wide (**Figure 62**). The reef is composed of dead reef rock with living corals abundant on the seaward side. The seaward slope drops off rapidly from the narrow reef or platform. Wave action, including breakers, are constantly breaking apart the seaward side of the reef.

Fractured rock masses are broken off from the edge and are washed up onto the shore behind the exposed slope. Most of the fragments fall down slope forming a seaward talus slope. The inner reef consists of a boulder zone where reef fragments and an inner flat are present.

Between the boulder zone, island, or continental mass where the reef itself is attached, there is an inner flat consisting of a small flat channel where growth and decay balance out. When the reef grows seaward, this inner channel widens and deepens merging into a barrier reef structure. The barrier forms breakwater which protects the inner lagoon. Near the edges of the lagoon, lime sand covers the bottom of the lagoon. Large branching corals occupy quiet water habitat.

In the deeper part of the lagoon, coral sand grades into mud. Calcareous sediments and corals are exposed to solution and mud particles are removed by suspension. Evaporites, oolites, red and green shale forms in lagoonal settings. Atolls, or table reefs, are either partially submerged or partially emergent.

## Fluid Traps

Differences in hydraulic gradients within an aquifer which controls the flow of the ground water down slope may prevent upslope movements of petroleum within the aquifer. Due to the buoyancy of petroleum, a pool will developed (**Figure 63**).

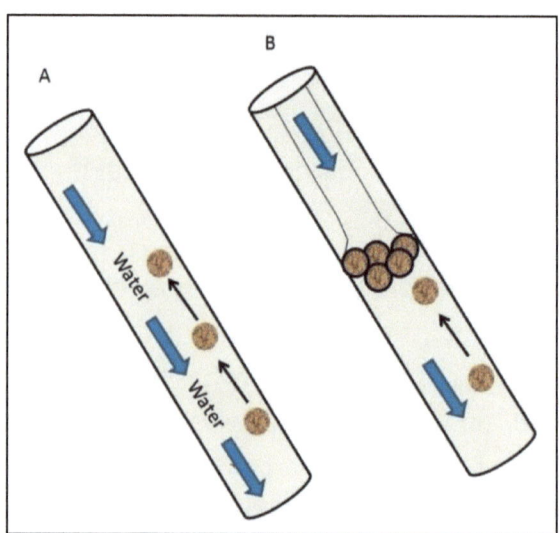

*Figure 63. Tube A ground water flows downslope. Petroleum droplets have enough buoyancy to flow upwards against the ground water pressure gradient. In Tube B, a constriction in the tube increases the ground water pressure gradient and velocity in the down slope direction. The increases create a localized downward force greater than the buoyancy effect, resulting in petroleum droplets accumulating at the constriction. Source: Redrawn from Levorsen, 1967.*

## Combination Traps

Combination traps combine structural, stratigraphic, and fluid traps in varying proportions. These types of traps have a two or three development stage history: 1) stratigraphy caused reduced permeability in the reservoir rock; 2) structure was developed by deformation which combines the stratigraphic element to form the rock portion of the trap; 3) a down dip ground water pressure gradient increasing the trapping effect. Where large pools accumulate, petroleum is trapped by ground water pressure gradients, which in turn, helps the geologic trapping mechanisms to create a barrier zone. Reduced permeability in the barrier zone may result from stratigraphic variations or from structural deformation or faulting which influence ground water flow gradients.

The Bighorn Basin in Wyoming is characterized by predominantly structural traps with secondary stratigraphic trapping within reservoir rock formations capped by low permeability strata. Chapter 5 presents the geologic structure of the basin along with petroleum trapping mechanisms.

Thom (1952) divided the deformational history of the Yellowstone-Bighorn region into seven main tectonic phases. During Early Precambrian or Archaeozoic phase, mountain building and geosynclinal sedimentation was followed by mountain uplift and compression. Intrusions in the form of batholiths, lopoliths (a large floored intrusion whereby the center is sunken into a basin like form), and dikes occurred. Regional fracture systems subdivided basement rocks into separate, individually recognizable blocks.

During the second phase, a long period of crustal stability was followed by erosion which truncated many formations resulting in development of lowlands supported by beveled remnants of worn down mountain ranges deformed during the first phase.

Between late Precambrian Era and Permian Period, western Montana and Idaho developed a geosyncline on top of eroded Precambrian basement. Large volumes of sediments collected within the geosynclinal structure. Shallow seas invaded and spread across the northern US into Minnesota. Small islands and peninsulas developed in shallow seas in response to vertical readjustments along one or several Precambrian rifts.

The fourth phase belonged to the Nevadan-Laramide Uplift. Triassic deformation, compression, volcanic pulses, and batholithic intrusions occurred near the Pacific Coast. Between Jurassic and Cretaceous time, Laramide compression and thrusting shifted eastward until it reached the Front Ranges of the Rocky Mountains at the end of the Paleocene Epoch.

Following Laramide uplift, a readjustment phase within the Cordilleran region became depressed although antecedent linear ridges continued to bow upwards under thrust faulted compression. Deep valleys were eroded by rivers between the uplifting ridges during the Eocene Epoch (**Figure 64**).

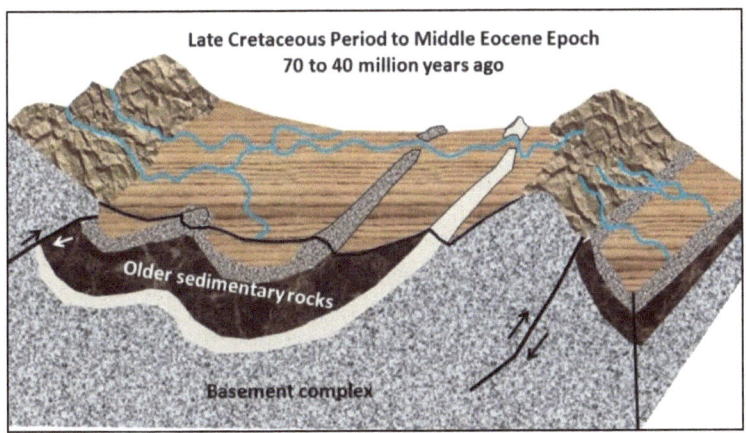

*Figure 64. Mountains and basins formed by folding and faulting. Extensive erosion began. Source: Redrawn from Lageson and Spearing, 1988, adapted from S.H. Knight, Univ. of Wyoming.*

Volcanic eruptions began in the Yellowstone region and intermontane basins began to accumulate lake and playa sedimentary sequences. During the Oligocene and Miocene, mountain ridges began to erode large volumes of sediments along with volcanic ash and dust, filling in the intermontane basins. Thick deposits began to cover over the ridge crests burying the mountain ranges (**Figure 65**).

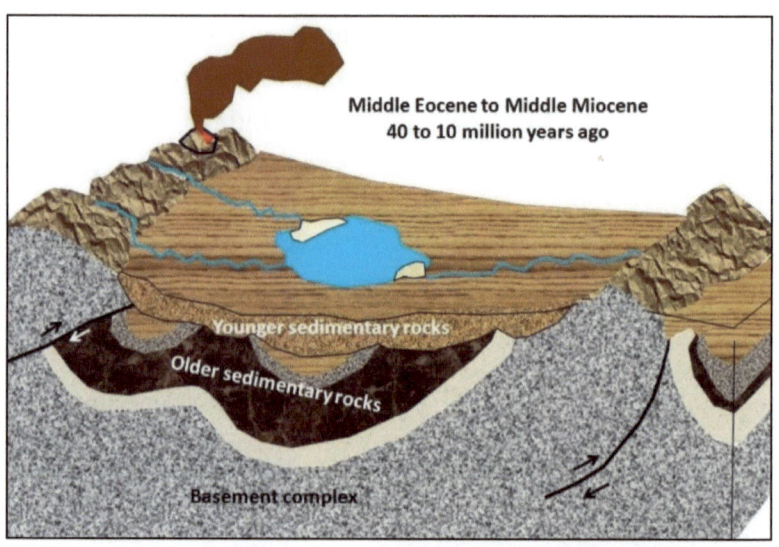

*Figure 65. Basins became partly filled by sediment shed from eroding mountains. Volcanic ash eruptions added to valley fill. Redrawn from Lageson and Spearing, 1988, adapted from S.H. Knight, Univ. of Wyoming.*

The sixth phase of Miocene-Pliocene Rocky Mountain evolution involved cut and fill processes which reduced the landscape to near sea level. These surfaces were called equilibrium surfaces, or more aptly, mature landscapes using geomorphology terminology. Basin valley fill and basement rock ridges became truncated by deposits left behind by meandering rivers. Pediment surfaces were left behind (**Figure 66**).

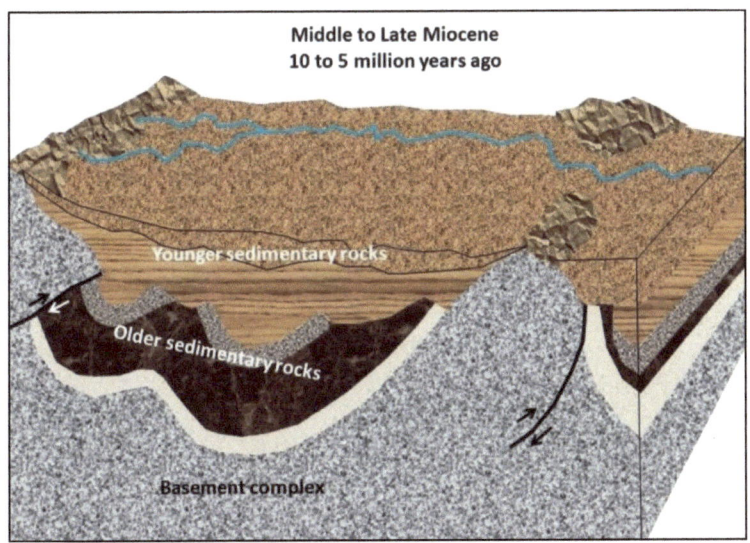

*Figure 66. Basins continued receiving sediments shed from eroding mountain ranges covering older sedimentary rock. Primary streams transected low passes through buried ranges. Redrawn from Lageson and Spearing, 1988, adapted from S.H. Knight, Univ. of Wyoming.*

The last phase consisted of Pliocene-Pleistocene arching of the Great Plains and eastern Cordilleran region. Arching resulted in development of Great Basin and Western Montana fault block systems. Re-excavation of valley fill by river entrenchment formed deep canyons. Stream capturing and glaciation of re-emerging ridges were commonplace (**Figure 67**).

## Tectonic Events

Compressional forces were applied perpendicular to the Rocky Mountain Cordilleran, generally in an ENE to WSW direction. Three first order tectonic features were developed: the North American continental shield or crustal plate on the east; the North Pacific sub-oceanic shield on the west; and, the Cordilleran zone occupying the central region between the continental shield and oceanic plate systems.

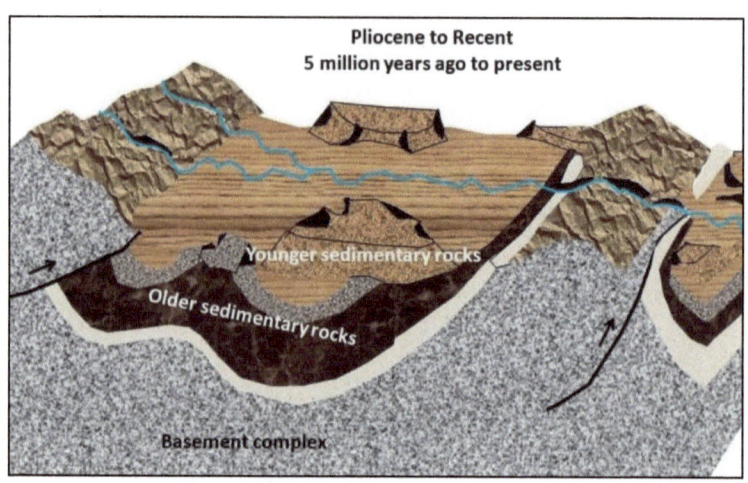

*Figure 67. Regional uplift rejuvenated new cycles of erosion, down cutting into valley fill sediments and exposing older buried sedimentary rocks. Canyons were excavated across buried mountain ranges when the ranges were exhumed. Redrawn from Lageson and Spearing, 1988, adapted from S.H. Knight, Univ. of Wyoming.*

Second order tectonic features included Laramide compression, flexing of thick, stiff geosyncline deposits of northwestern Montana; a fault range pattern developed in association with the eastern Rocky Mountain and Colorado Plateau Provinces; and, batholithic intrusions and arcuate thrusting of the wedge shaped area in Montana positioned between the Osburn Lake Basin and Jefferson Canyon-Ross Peak rift zones.

Third order tectonic features are recognized by crustal blocks formed by regionally compressive stresses defined by Precambrian rifting and intrusion.

Crustal plates were domed by Laramide intrusion or uplifted and tilted during Laramide compression; locally, downward or upward pointing of crustal wedges became elevated or depressed by compressional thrusting; and, larger sub basin crustal plates were forced to flex downward into basins by regionally applied compressional stresses.

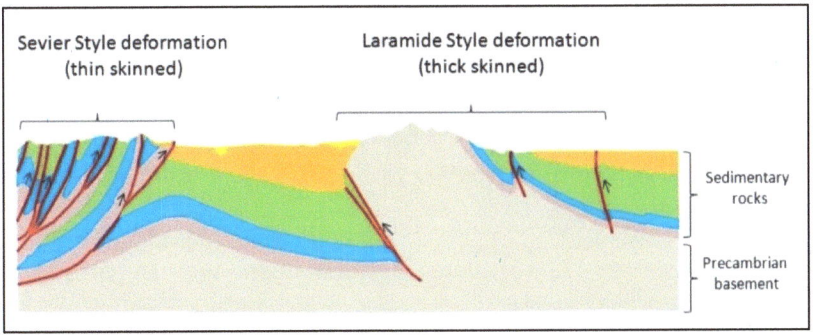

*Figure 68. Both Sevier and Laramide uplifts impacted Wyoming during Late Mesozoic into Early Cenozoic time. Laramide uplifting continued beyond the end of Sevier uplift. Source: Wyoming Geological Survey.*

## Sevier Tectonics

Between 140 and 50 million years ago, the Farallon oceanic plate was subducting beneath the western North American continental margin near California. Compression between the plates resulted in crustal thickening, uplifting portions of Nevada, Utah, and Wyoming. North to south trending mountain ranges were uplifted in the western Wyoming Thrust Belt, Overthrust Belt, or fold and thrust belt. Sevier mountains are west dipping with eastward convergent thrusts which displaced sedimentary rock above crystalline basement. Thrusting did not involve uplift of basement rock. This condition is referred to as thin skinned tectonics. Western Wyoming thrust belt ranges consist of the Snake River, Hoback, Salt River, Tunip, and Sublette Ranges.

## Laramide Tectonics

Overlapping the Sevier mountain uplift, the Laramide uplift began 70 million years ago and ended 35 million years ago. Horizontal compression began when the Farallon plate descended beneath the North American plate.

The Farallon plate changed subduction angle, flattening out beneath the continental margin, reaching further inland than the Sevier event. Crystalline basement rocks and the overlying sedimentary rocks were uplifted along faults. This process was referred to as thick skinned tectonics. Crustal thickening in the direction of subduction occurred in response to compression forces between converging plate margins (Wyoming State Geological Survey, 2016). The Bighorn Basin was deformed by Laramide type deformation.

*Western Basin Margin Folds*

Rattlesnake Mountain formed when a thrust fault was displaced along the western flank of the mountain which arched Paleozoic and Mesozoic rocks on top of Precambrian basement. Pennsylvanian Tensleep sandstone, Mississippian Madison limestone, and Ordovician Bighorn dolomite make up the mountain (**Figure 69**).

*Figure 69. Western flank of Rattlesnake Mountain represents steeply tilted sedimentary rocks belonging to Bighorn Dolomite, Madison Limestone, and Tensleep formation sandstone. Layers were tilted when Precambrian basement rocks were uplifted along thrust faults during the Laramide uplift.*

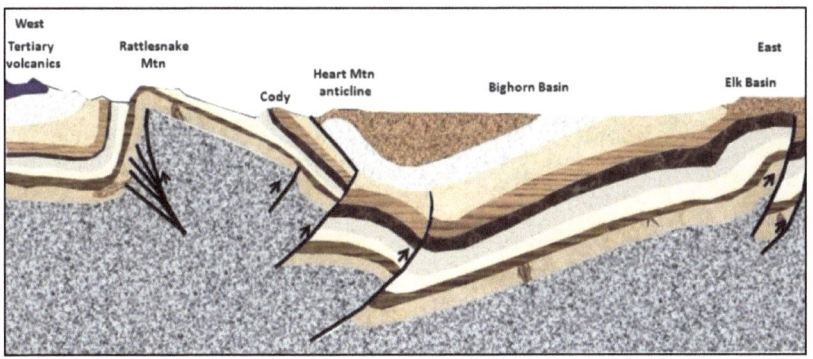

*Figure 70. Generalized cross section through the north central Bighorn Basin demonstrating thick skinned deformation of buried basement and overlying sedimentary rocks due to Laramide faulting and folding. Source: Redrawn from Lageson and Spearing, 1992.*

Archaen basement rocks were brought to the surface during the Laramide Uplift, beginning 70 million years ago. Most notably, the Pryor Mountains and other ranges previously buried during the Paleozoic and earlier Mesozoic Eras. Thrust faulting and folding of former horizontal sedimentary rocks deposited during both eras in the Bighorn Mountains and Basin occurred during deformation in a foreland thrust environment. The central part of the Bighorn Basin between Greybull and Cody occupy a regional synclinal structure (**Figure 70**).

The Bighorn Basin is part of Zone 2 (previously described in Chapter 3). Zone 2 consists of a platform composed of Mesozoic aged sedimentary rocks composed of anticlines and synclines which encircles the basin. The platform forms a bench between the high mountain ranges and flat basin. During Laramide uplift, the basin subsided. Displacement developed a zone of faults which uplifted the platform. Reverse and/or thrust faults bound the platform, buried deeply, providing the mechanism for raising Precambrian basement rock.

When the basement uplifted, overlying sedimentary rocks were folded. Anticlines are recognized by the bright red Triassic Chugwater formation, exposed along fold cores and flanks.

Cretaceous rocks, belonging to the Cody and Frontier formations, were tilted westward during the early stages of Laramide uplift. Erosion truncated the upper surfaces when the Paleocene Fort Union sediments buried Cretaceous rocks. Eocene Willwood formation claystone covered the Fort Union formation in the central basin when basin subsidence and mountain uplift tilted the basin margins.

The Heart Mountain anticline represents a regional detachment fault where large blocks of Paleozoic strata slid from the Beartooth Mountains eastward on top of the Willwood formation. Sliding began about 50 to 45 million years ago by gravity, similar in displacement to landslide movements. Earthquakes, fluid lubrication, or volcanic gases provided the slide plane surface for displacement to occur by tens of miles without disrupting sedimentary stratigraphic sequencing.

The eastern edge of the basin belongs to Zone 1, called the Basin Margin anticlinal zone. The Elk Basin anticline represents a series of faulted and folded anticlines and domes which developed during the Laramide uplift. Basin margin anticlines formed during Laramide faulting and folding events around the Bighorn basin margins. Uplifting of Precambrian basement rocks arched the overlying sedimentary rock column above grade, allowing erosion to strip rocks younger than the Cretaceous Cody formation from the western flanks of the Horse Creek anticline, leaving Frontier and Dakota sandstone exposed in the central fold axis region (**Figure 71**). The Oregon Basin dome is separated from the Horse Creek anticline by a regional syncline.

The syncline formed when subsurface thrusting down dropped Paleozoic through Cretaceous formation rocks beneath the Mesaverde Formation. The adjacent anticline to the east became the Oregon Basin dome, arching up the Mesaverde Formation resulting in the erosion of the younger overlying Meeteetsie and Fort Union formations. The basin obtained its name from the Mesaverde formation outer rim structure leaving behind a depressed inner axial region. The Bighorn Basin makes up the remaining central and eastern section of Figure 70, accumulating thick sedimentary sequences from sediments shed during the western regional Laramide uplift.

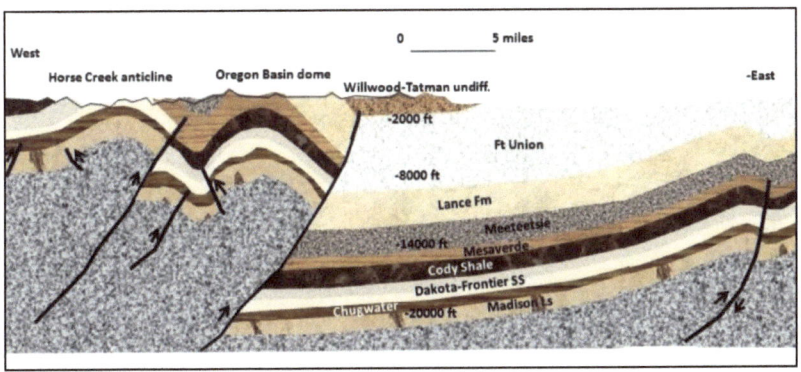

*Figure 71. South of the Rattlesnake Mountain section, the west side of the basin margin anticlinal zone passes through the Oregon Basin dome into the eastern Bighorn Basin axis. Source: Redrawn from Lageson and Spearing, 1992.*

South of Cody, Wyoming along the western side of WY Highway 120, the Horse Creek Anticline marks the central position of the basin margin anticlinal zone. Arching of the basement complex uplifted the overlying Paleozoic and Mesozoic sedimentary rocks exposing the Dakota, Frontier, and Cody formations to erosion at the surface. East of the Horse Creek Anticline, near Elk Butte, the Oregon Basin dome was arched upwards along a set of westward dipping thrust faults extending through Mesozoic and Paleozoic strata.

Eastward dipping faulting occurred within the Early Cretaceous Dakota and Frontier formations. Faulting helped uplift the Oregon basin dome structure. The Bighorn Basin received thick sedimentary deposits shed from rising anticlinal structures to the west, separated by a large westward dipping thrust fault. Along the eastern basin margin, basement thrusting arched buried strata beneath the Fort Union Formation.

*Figure 72. South of the Oregon Basin dome, the Pitchfork, Spring Creek Rawhide, and Meeteetsie Anticlines make up the basin margin anticlinal zone.*

The Pitchfork anticline plunges at Wyoming Highway 290, 1.5 miles west of the town of Meeteetsie. The anticline was formed from folding occurring on the up-thrown side of the Pitchfork thrust fault. Precambrian basement displaced the overlying Madison, Chugwater, and Frontier Formations. The Cody Formation and younger sedimentary rocks were eroded from the surface during uplift (**Figure 72**).

The Spring Creek - Rawhide anticline intersects Wyoming Highway 290, approximately 0.75 miles west of Meeteetsie, east of the Pitchfork anticline. Spring Creek folding was formed in a similar manner by thrust faulting associated with the up-thrown side of the Spring Creek thrust.

The Meeteetsie Anticline is located northeast of the town, approximately 1 mile east of Wyoming 120. Thick deposits accumulated in the central part of the section syncline, later interrupted by uplift of the Meeteetsie anticline along a set of thrust faults in the northwestern part of the basin. The left thrust of the Meeteetsie anticline developed first in response to the same compressive forces which uplifted both folds in the southeastern part of the section. The right thrust developed later, cutting across the lower end of the left fault, possibly responding to Beartooth-Rattlesnake Mountain uplifting.

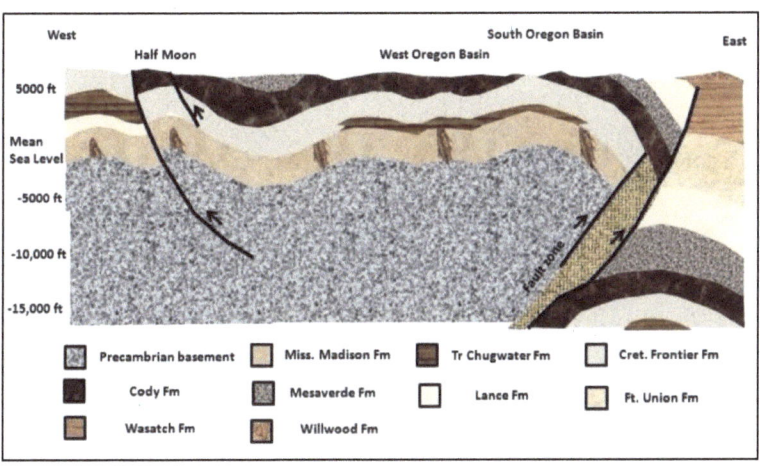

*Figure 73. Further south, folding appears less intense than the previous sections. Half Moon, West and South Oregon Basins represent the basin margin anticlinal zone. More intense folding appears east of the South Oregon Basin structure due to a buried fault zone.*

The Half Moon anticline, located approximately 1 mile west of Wyoming Highway 120, is positioned in the shallow basin margin deformation zone. This zone was formed by folding associated with two parallel thrust faults which arched up into a central fold axial plane. The right fault plane influenced development of the synclinal fold positioned along the left flank of West Oregon Basin. The West Oregon basin intersects with Wyoming 120, west of the main Oregon dome structure.

The South Oregon Basin dome was arched upwards in response to a subsurface fault zone buried beneath the Cody formation along the eastern edge of the basin. Younger sedimentary rocks occurring above the Cody shale were eroded from the basin during uplift. South Oregon Basin is located northeast of the intersection between Wyoming Highways 120 and 30. Thick deposits were shed eastward into a subsiding basin, later folded into an anticline and uplifted along the right thrust fault. Wasatch formation sediments were eroded and covered by the Willwood formation (**Figure 73**).

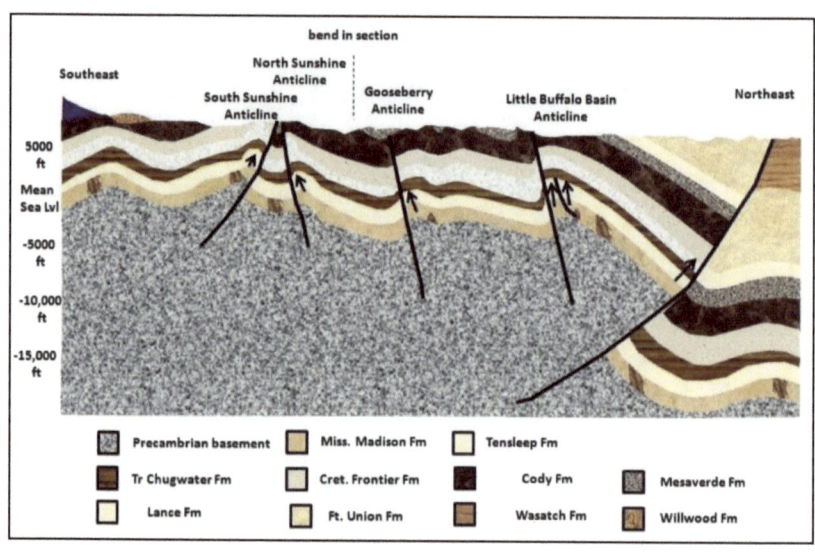

*Figure 74. South of Half Moon, deformation was more intense when smaller thrust faults become more numerous and more closely spaced together.*

South Sunshine Anticline is located in the central western basin margin deformation zone, positioned approximately 0.5 miles southwest of North Sunshine Anticline (**Figure 74**). The fold was arched upwards by a southeastward dipping thrust fault forcing the Frontier formation to the surface where it was eroded. Arching caused the area to the southeast to become deformed into a syncline-anticlinal folded sequence.

A second thrust fault, dipping to the northeast, pinched a steep synclinal basin between the South and North Sunshine folds. The same thrust also arched up the North Sunshine anticline, pushing the Cody Formation to the surface where it was eroded. Eastward, the Cody formation remains exposed at the surface where a synclinal basin developed leading up the next northeastward dipping thrust. The Gooseberry anticline was deformed by thrusting where a regional anticlinal arch developed smaller anticlinal and synclinal features distorted by the Little Buffalo northeast dipping thrust fault. The Cody, Mesaverde, Frontier, and younger Cenozoic strata were removed by erosion east of the Buffalo fold. The Bighorn Basin axial thrust became exposed at grade by erosion of the Wasatch Formation from the northeastern edge of the Bighorn Basin.

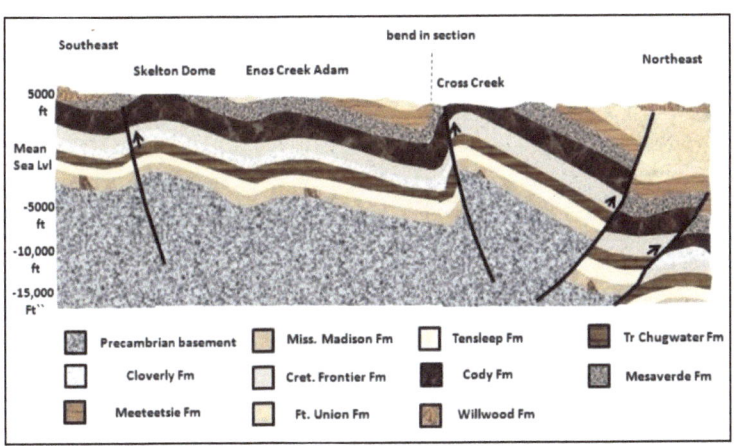

*Figure 75. Further south of Half Moon, Skelton dome and Enos Creek are subdued folds. Northeastward, Grass Creek and the Bighorn Basin are more intensely folded.*

Skelton Dome is positioned very close to the Absaroka Volcanic plateau along the western basin margin, approximately 1 mile south of the Enos Creek anticline (**Figure 75**). A northeast dipping thrust fault extending from the basement into the Mesozoic strata pushed up Skelton Dome until the overlying Mesaverde formation and the upper parts of the Cody formation were eroded from the landscape. Faulting also influenced uplift of Enos Creek anticline north of Skelton. A bend in the cross section crosses the Grass Creek thrust fault, dipping to the northeast which played a more intense uplift role for the Grass Creek anticline. Uplift stripped away Mesaverde and Cody formation sedimentary rocks. To the northeast, a pair of southeast dipping thrust faults deformed Mesozoic, Paleozoic, and Precambrian basement rocks buried by Lance Formation sediments.

## *Southern Basin Marginal Folds*

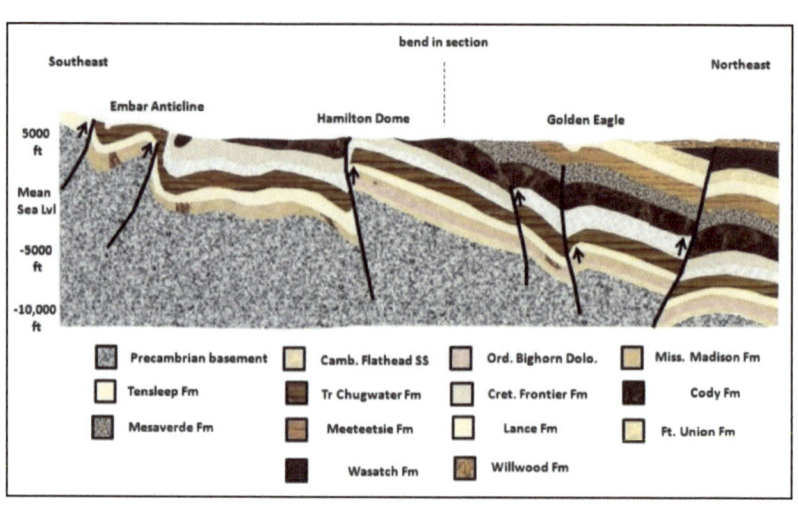

*Figure 76. Southern section of the basin margin deformation zone includes a moderately intense anticline fold, buried synclinal basin, followed by a steeply deformed dome and buried folded sequences to the northeast.*

Within the western part of the Wind River Indian Reservation, approximately 0.5 mile northeast of the Owl Creek Fault, the Embar anticline was warped by a set of southeastward dipping thrusts spaced relatively close together (**Figure 76**). A steeply dipping synclinal fold and adjacent anticlinal fold makes up the Embar Anticlinal structure. Folding eroded much of the Mesozoic strata from above the Triassic Chugwater formation which is exposed at the surface. To the northeast, the second thrust down warped a synclinal basin which removed most of the Cody Formation, exposing the underlying Frontier Formation. Buried beneath the Cody formation, a set of warped anticlines and synclines remained buried consisting of Cloverly, Chugwater, Tensleep, and Madison formation sediments. The steeply dipping Hamilton dome thrust (northeast dip) uplifted the Hamilton Dome structure. The dome is located on Wyoming 170, at the northern edge of the Wind River Indian Reservation. Most of the Frontier formation was eroded from the dome, exposing Cloverly formation sandstone.

Sedimentary rocks dip to the northeast leading up to a set of northeasterly dipping thrusts which warped the region between Hamilton Dome and the Golden Eagle anticline. The western part of the Bighorn Basin was arched into an anticlinal structure, buried by Cody and Willwood formations.

A buried thrust arched the Golden Eagle anticline covered by the Wasatch formation. Wasatch formation and younger rocks are exposed to the east of the fold leading into the Bighorn Basin thrust.

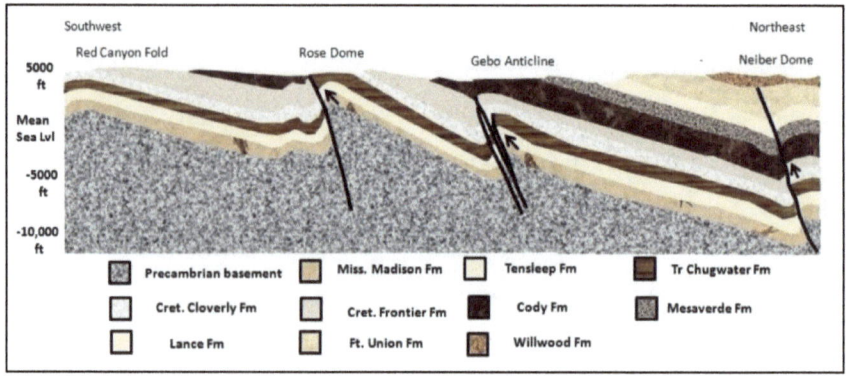

*Figure 77. The southern basin margin fold zone is focused about the Wind River Indian Reservation extending northeast towards Thermopolis, Wyoming. Deformation intensity increased towards the northeast reaching its maximum in the center region.*

Approximately 1 mile west of Thermopolis, Wyoming, south of Wyoming 170 intersection with Wyoming 120, the Red Canyon Fold was uplifted along the steeply dipping northeast Red Canyon Thrust (not shown) (**Figure 77**). Arching of the basement pushed the Frontier-Cody and younger formation strata upwards to the surface where erosion stripped the sedimentary rocks from grade. To the east, a synclinal basin developed west of the Rose Dome thrust. Older Mesozoic Chugwater formation rocks and the overlying formations were eroded from the upper fold axial region of the dome structure. Between Rose Dome and Gebo anticline, steeply dipping thrust faults to the northeast intensified folding of both the dome and anticlinal structures.

A steeply defined syncline developed southwest of Gebo in response to fault splintering below the Cody Formation. Frontier and younger formations were eroded from the anticlinal uplift. Most of the buried deformation was focused in the early Mesozoic and older aged rock formations. Northeast of Gebo, deformation was less intense approaching the Neiber Thrust and Dome sequence. The dome was mostly deformed in older rocks buried beneath the Willwood formation unconformity.

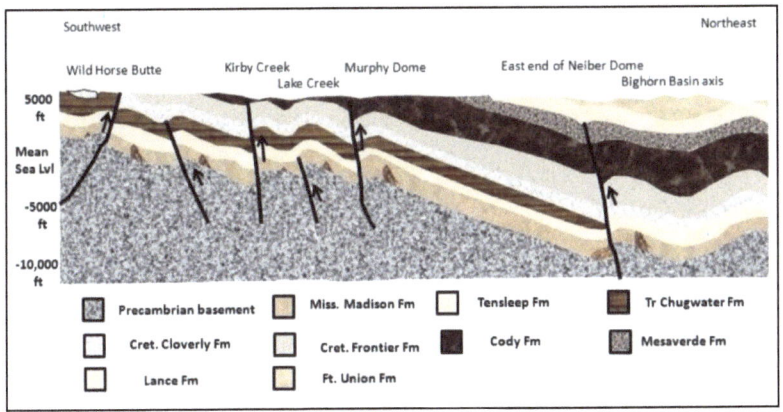

*Figure 78. Intense deformation occurred within the southeastern basin region, producing closely spaced fold structures east of Warm Springs, Wyoming.*

The southeast end of the Bighorn Basin was intensely deformed by a series of closely spaced anticlines, synclines, and domal folds (**Figure 78**). Thrust faults are nearly vertical in orientation, tilted slightly to the northeast. The exception belongs to the Wild Horse Butte thrust which tilts to the southeast direction. Folding eroded the upper Frontier formation, overlying Cody formation and younger rocks between Wild Horse Butte and Kirby Creek dome. A synclinal fold occurring between Wild Horse Butte and Kirby Creek was warped into a central anticline in the subsurface by a buried thrust which terminated into the Chugwater Formation. The fold was covered over by Frontier formation sedimentation. Most of the upper part of the Frontier and all of the younger strata were eroded by the uplift.

Kirby Creek, Lake Creek, and Murphy Dome were more intensely deformed into tightly folded structures by a series of near vertically dipping thrust faults. The center fault pushed up the Lake Creek anticline from a subsurface zone, warping the overlying strata into a deformed folded sequence in the overlying rocks. Remnants of the Cody Formation were left exposed at the surface.

Northeast from Murphy Dome, formations descend into the Bighorn Basin which was deformed by a northeast dipping thrust, buried by the Mesaverde Formation. The fault deformed the basin subsurface into a broad anticline-syncline-anticline sequence. The Fort Union formation is exposed in the central basin surrounded by Lance Formation.

The Bighorn Basin belongs to the Rocky Mountain foreland exposed to late Laramide tectonism. Petroleum discovery and development occurred within the basin margin deformation belt rimming the outer zone within the basin. Both structural and stratigraphic trapping mechanisms were responsible for accumulating pools in Permian and Cretaceous source rock and reservoir systems.

### Conventional Plays

Conventional plays were defined by Fox and Dalton (no published date) of the USGS with help from state geological surveys. The Basin Margin Subthrust, Basin Margin Anticline, Deep Basin Structure, Sub Absaroka, Phosphoria Stratigraphic, Tensleep Paleotopographic, Greybull-Cloverly-Muddy Sandstone Stratigraphic, Bighorn-Darby Wedge-Edge Pinchout, Flathead-Lander and Equivalent Sandstone Stratigraphic, Madison Limestone Stratigraphic, Darwin-Amsden Sandstone Stratigraphic, Triassic-Jurassic Stratigraphic, Cody and Frontier Stratigraphic, and the Shallow Tertiary-Upper Cretaceous Stratigraphic Plays are the most common provinces were conventional petroleum reservoirs are located. One unconventional play was noted: Basin Center Gas Play. Significant petroleum production was found in the Basin Margin, Deep Basin, and Phosphoria Stratigraphic plays.

### Basin Margin Subthrust Play

Laramide basin margin thrusting is thought to be the primary petroleum trapping mechanism in upturned, overturned, folded and faulted Phanerozoic strata below the overthrust wedge. The basin mostly occurs within the Wind River basin and within southwestern Wyoming.

Porous and permeable sandstone possess good reservoir quality along with fracturing associated with thrust faulting.

*Reservoirs.* Principle reservoirs belong to the Pennsylvanian Tensleep formation sandstone, Permian Phosphoria limestone, and Upper Cretaceous Frontier Sandstone. Source rocks belong to Permian Phosphoria formation carbonates, and Cretaceous Mowry, Frontier, Mesaverde, and Meeteesie formation rocks.

*Tectonic Mechanisms.* Laramide uplift brought thick wedges of Precambrian basement rock on top of Phanerozoic rocks by thrust faulting. Hydrocarbons from deep source rocks migrated from source areas buried deep within the basin during or after deformation. Theories propose pre-Laramide migration may have occurred from the Phosphoria Formation, moving hydrocarbons into reservoirs prior to tectonic development of basin marginal faults and folds. Stratigraphic traps may have formed prior to deformation followed by superimposed structural trapping post stratigraphic trapping.

*Trapping Mechanisms.* Structural closures occurring below thrust faults may be sealed by impermeable rocks in the hanging wall. During the thrusting process, underlying beds were folded, often upturned or overturned resulting in slicing faults forming segmented slivers. Petroleum became trapped within the upturned and overturned folds and faulted strata. Production depths may occur up to 20,000 feet on the steeper side of the asymmetrical basin produced by folding to less than 10,000 feet in other basin margin zones.

Petroleum became trapped in faulted anticlines and domes, and within faulted fold noses that formed during the Laramide uplift. Structures are most pronounced along the shallow basin margins with production occurring from a few thousand feet up to 12,000 feet.

*Reservoirs.* Producing formations occur within Cambrian to Late Cretaceous aged formations including the Flathead, Bighorn, Jefferson, Madison, Amsden, Tensleep, Phosphoria, Dinwoody, Crow Mountain, Chugwater, Cloverly, Dakota, Greybull, Lakota, Muddy, Frontier, and Mesaverde Formations. Primary production occurs from the Madison, Tensleep, Phosphoria, and Frontier formations. Formations exhibit common oil-water contact zones for the Paleozoic reservoirs. Most reservoirs are less than 50 feet thick.

Source rocks consist of organically rich clayey sedimentary rocks belonging to two distinct geochemically identical groups: Permian and Cretaceous aged formations. Permian source rocks occur within the Phosphoria Formation. Cretaceous source rocks occur within the Mowry, Frontier, Mesaverde, and Meeteetsie Formations. Deeply buried rocks are exposed to higher thermal gradients which promotes a predominance of gas deposits.

*Timing Mechanisms.* Local migration of oil without long distance transport occurred during the Laramide Uplift. A second theory proposes oil formation occurred prior to the Laramide Uplift coupled with long distance migration from Western Wyoming before basin formation occurred with remigration influenced by the Laramide event. Cretaceous source rocks are known to have reached maturity during the early Paleocene Epoch in the deeper parts of the basin followed by development within the younger rocks.

During the Laramide event, structural uplift coincided with migration maturity which influenced the final concentration process.

*Trapping Mechanisms.* Anticlines and domes, most of which are faulted provide the closure mechanism for petroleum accumulation. Faulted noses promote trapping, as well. Structures are more pronounced around the shallow margins of the basin. Production usually occurs between a few hundred feet up to 12,000 feet deep. Within these fold structures, impermeable beds act as seals which separate Paleozoic from Mesozoic reservoirs.

### Deep Basin Structural Play

Within the deeper portions of the central Bighorn Basin, large anticlinal and fold nose structures accumulated liquefied natural gas, and vapor gas deposits. Reservoirs cut across the basin axis diagonally. Example reservoirs include the Five Mile-Dobie Creek Field.

*Reservoirs.* Permian to Cretaceous formations make up the reservoir rocks including the Phosphoria, Tensleep, Muddy, and Frontier Formations. Higher porosities occur within the Frontier and Muddy sandstones within the Five Mile-Dobie Creek fields. Porosity is limited by burial diagenesis. Carbonate reservoirs in the Phosphoria Formation are most productive. Tensleep Sandstone has not proven to be of commercial value but hydrocarbon shows suggest a potential exists, limited to deep burial. Reservoirs range between 25 and 100 feet thick.

*Source Rocks.* Phosphoria Formation marine shale and swamp shale belonging to the Cretaceous Thermopolis, Mowry, Frontier, Cody, and Mesaverde Formations are the primary sources for petroleum.

Most of the formations are deeply buried, possibly beyond the maturity range for oil and gas generation including the gas zone.

*Timing Mechanism.* Deep burial of Cretaceous shale and the Phosphoria formation in the western part of the basin may have allowed petroleum migration to begin before Laramide uplift during Late Cretaceous time. Both Permian and Cretaceous source rocks may be present up to 15,000 feet deep or greater in the northwest part of the deep basin. Cretaceous sources reached maturity by the early Paleocene Epoch. Reservoir sandstones in the Frontier and Mesaverde formations are interbedded with marine source rocks. Migration occurs between the source and reservoir rocks under these conditions.

*Trapping Mechanisms.* Primary traps belong to Laramide intra-basin anticlines which are faulted on the northern side, referred to as the Five Mile Trend. The trend is oriented diagonally in a northwest direction across the basin center, the only large structure in the play zone. Plunges occur to the northwest with depth to the top of the Tensleep formation at 25,000 feet, at the northwest end. At the southeastern end, near the Cottonwood Creek, Worland, and Rattlesnake Creek fields, the Tensleep formation shallows to 11,000 feet. Reservoirs are sealed by fine grained beds interbedded with reservoir rocks which formerly served as source rocks. Potential production occurs between 10,000 on the southeastern side and 20,000 feet on the northwestern side of the basin.

### Sub Absaroka Play

Oil occurs beneath Eocene aged volcanic rocks, trapped by Laramide anticlines and domes. May folds were faulted and eroded prior to volcanic coverage. The eastern part of the province is defined by the presence of overlying volcanics except for the Beartooth Uplift region.

The northern limit is defined by contact with Precambrian basement rock. The western limit is defined by truncated subcropping strata near the Yellowstone boundary. The southern limit is defined by the southern extent of Eocene volcanics.

*Reservoirs.* Quartz sandstone is the primary production unit from the Pennsylvanian Tensleep Sandstone. Secondary production occurs from the carbonate Mississippian Madison Limestone, Permian Phosphoria and Dinwoody formations, and sandstones of the Pennsylvanian Darwin, Triassic Chugwater, Jurassic Curtis, and the Lower and Upper Cretaceous formations. In some locations, secondary cementation has destroyed primary porosity.

Within the Tensleep Formation, reservoir porosity is very high at shallow depths. Moderate porosity is present in the Madison, Amsden, Phosphoria, and Dinwoody formations. Thickness varies by geologic unit: Phosphoria production thicknesses are 35 feet; Tensleep Sandstone 75 feet; Chugwater Formation thickness is 50 feet; and Frontier Formation production is 40 feet thick.

*Source Rocks.* Carbon rich source rocks exist in the Paleozoic and Mesozoic formations within the basin. Many of these rocks were deeply buried enough to generate hydrocarbons prior to Laramide uplift.

*Timing Mechanisms.* The earliest formation of hydrocarbons occurred in the eastern side of the play zone, near the basin axis. Faulting was severe in the western side, resulting from Laramide uplifting. These faults provided conduits for hydrocarbon migration vertically upward until porous and permeable reservoir rocks were reached and trapped. Long distance migration of Phosphoria source oil from the west probably occurred before basin formation with remigration occurring during the Laramide uplift.

*Trapping Mechanisms.* Domes and plunging anticlines formed trapping mechanisms for hydrocarbons. These traps combined with faulted anticlinal structures. Structures within the eastern play occur along the trend of the basin margin anticlinal play. Impermeable beds within these structures form the trapping seals. Multiple play zones may occur, considering the extensive faulting of the zone. Oil seeps are numerous along fractures in the overlying volcanics. Depths are difficult to predict based on the rigged nature of the Absaroka Mountains.

### Phosphoria Stratigraphic Play

The Phosphoria Ervay Member contains high sulfur oil within stratigraphic traps trending in a north to south direction. Carbonates transition from the west into red shale and evaporites belonging to the Goose Egg formation in the east. Production is centered in the eastern Bighorn Basin bounded by the Ervay Tongue and in the west by the down dip limit of oil accumulation. The north and south boundaries are defined by Phosphoria formation exposures.

*Reservoirs.* Reservoirs occur within the Permian Ervay Member of the Phosphoria Formation. The Ervay Member consists of dolomitized grainstones and packstones. Algal rocks may locally contribute to petroleum accumulations. Depositional settings consist of high energy tidal and associated environments. The Cottonwood Creek field was formed by high energy tidal channels sealed in the updip direction by fine grained inter- and supra- tidal carbonates. Fractures enhance formation porosities. Thicknesses range from 25 to 75 feet.

*Source Rocks.* Organic rich Permian Phosphoria shale source rocks in the western part of the basin were deeply buried, promoting hydrocarbon generation.

*Timing Mechanisms.* A combination of Laramide related and pre-Laramide oil generation and migration possibly occurred. Phosphoria source rocks possibly began as early as Jurassic time in western Wyoming and eastern Idaho.

*Trapping Mechanisms.* Stratigraphic traps occur near the Ervay Member carbonate tongue in porous, detrital reservoirs deposited within high energy tidal flats on coastal plains sealed by updip, fine grained tight carbonates and red shale formed in both intra- and supra-tidal environments. Lateral traps formed in muddy carbonates within the Ervay member. Regionally, the trap becomes a facies change from carbonates into red bed shale.

Vertically, seals are fine grained rocks overlying Triassic Dinwoody and Chugwater Formations with internal foot seals provided by fine grained red beds and carbonates. Production depths are 12,000 feet.

## Hypothetical Plays

Hypothetical plays are defined by zones where a potential for oil and gas accumulation are expected to occur but have not been proven by exploration and/or development.

### Tensleep Paleo-topography Play

Oil and gas accumulations are controlled by paleo-topographic erosion which occurred on the top of the Tensleep Sandstone within areas of low structural relief within the Bighorn Basin.

*Reservoirs.* Eolian and beach facies consisting of fine to medium orthoquartzites with moderate to high porosities decrease with depth. Increasing depth effects inhibited loss of porosity during early hydrocarbon migration in filled reservoirs.

*Source Rocks.* Source rocks are restricted to organic rich shale belonging to the Phosphoria Formation.

*Timing Mechanisms.* Hydrocarbon migration is believed to have occurred before Laramide uplift and deformation.

*Trapping Mechanisms.* The upper surface of the Tensleep Formation is truncated by a regional unconformity. Paleo-topographic relief is widespread, occurring between 100 to 150 feet in relief. Stratigraphic traps appear in places where structural dips are less than paleo-slope relief which promotes remigration of hydrocarbons. Traps become modified by diagenesis and depositional influences associated with sandstone and dolomite. Seals are formed by Phosphoria formation dolomite and shale, or by Permian Goose Egg formation shale and evaporites where they overlie the Tensleep formation. Traps range in depth between 4000 to 13,000 feet. Below the lower limit, reservoirs tend to be tight.

*Resource Potential.* Resource potential is considered to be moderate with small accumulations. Production may occur from the upper Minnelusa formation in the Powder River Basin. Fields would be small but numerous.

*Greybull-Cloverly- Muddy Sandstone Stratigraphic Play*

Stratigraphic traps appear to be focused at the base of the Dakota Group (Greybull and Lakota sandstones). Lakota pool production is related to Laramide structures around the perimeter or within the play zone suggesting reservoir potential exists but does not represent actual sizes of the stratigraphic trap within the zone.

*Reservoirs.* Fine to medium sandstone, locally pebbly or conglomeratic containing abundant chert, appears to make up reservoir composition for this play zone. Oil accumulates within discrete channels or channel sandstone deposited under alluvial conditions. Production depths range between 4000 and 13,000 feet. Below 13,000 feet, reservoirs are expected to be tight.

*Source Rocks.* Cretaceous marine shale in the Mowry and Thermopolis formations make up the source rocks within the zone. Type II and III kerogen is the predominant petroleum product.

*Timing Mechanisms.* Mowry source rocks in some parts of the basin entered the thermal zone which influenced liquid hydrocarbon generation during the Paleocene, exiting from the thermal zone during the Miocene Epoch.

*Trapping Mechanisms.* Traps occur within alluvial channels common to combination structural noses or anticlinal closures. Major structural closures are not considered to be part of the play zone. Essential trapping mechanisms are stratigraphic. Seals are associated with fine grained alluvial and deltaic plain rocks usually found in the Fusion Shale or Jurassic Morrison Formation. Vertical seals belong to the overlying Thermopolis and Fusion Shale, or Dakota Sandstone. Lateral seals are formed by the Jurassic Morrison formation where it is entrenched by sandstone reservoirs.

*Resource Potential.* Undiscovered accumulations are estimated to be substantial with a few reaching appreciable size. Structural and combination traps appear to be the most promising for production. Stratigraphic accumulations have not been discovered since 1993. A large number of pools are estimated.

Bighorn Dolomite hydrocarbons appear to be associated with wedge-edge or bevealed edge pinch outs which abut against the base of the Madison Limestone. Hydrocarbon occurrences or source rocks are unknown.

*Reservoirs.* Reservoirs within the Bighorn Dolomite exhibit intergranular porosity within most of the play zone.

*Source Rocks.* No source rocks have been identified within the zone. This zone appears to be of high risk with low potential for successfully producing hydrocarbon accumulations.

*Timing and Trapping Mechanisms.* Timing mechanisms are unknown. The presence of trapping mechanisms at the Bighorn Dolomite-Madison Limestone unconformity has not produced demonstration of petroleum trapping.

*Resource Potential.* Petroleum exploration is placed at a very high risk for successfully developing hydrocarbons due to poor trap potential. Quantitative estimates of hydrocarbon accumulation could not be established.

### *Flathead-Lander and Equivalent Sandstone Stratigraphic Play*

Hydrocarbons are trapped within stratigraphic pinch outs between the Cambrian Flathead and Ordovician Lander Sandstones. Hydrocarbon occurrences or source rocks are unknown.

*Reservoirs.* Sandstone reservoirs cover much of the play zone and are considered to be widely variable in properties. Reservoir quality is considered poor based upon diagenetic variations.

*Source Rocks.* Source rocks are unknown resulting in a higher risk of hydrocarbon absence.

*Timing and Trapping Mechanisms.* Timing mechanisms are unknown. Stratigraphic pinch outs are anticipated but definite traps have not been identified.

*Resource Potential.* Petroleum exploration is placed at a very high risk of successfully developing hydrocarbons due to poor trap potential. Quantitative estimates of hydrocarbon accumulation could not be established.

### *Madison Limestone Stratigraphic Play.*

Hydrocarbon entrapment is thought to be associated with porosity variation and topographic features related to karst development.

*Reservoirs.* Vuggy karstic reservoirs are expected to cover the entire play zone in the upper part of the Madison Limestone

*Source Rocks.* Source rocks are unknown resulting in a higher risk of hydrocarbon absence.

*Timing and Trapping Mechanisms.* Timing mechanisms are unknown. Stratigraphic pinch outs are anticipated but definite traps have not been identified.

*Resource Potential.* Petroleum exploration is placed at a very high risk of successfully developing hydrocarbons due to poor trap potential. Quantitative estimates of hydrocarbon accumulation could not be established.

### *Darwin-Amsden Sandstone Stratigraphic Play*

Stratigraphic entrapment of oil in discontinuous sandstones of the Pennsylvanian Darwin and Amsden Formation are present in this play zone.

*Reservoirs.* Sandstone reservoirs cover much of the play zone and are considered to be widely variable in properties. Reservoir quality is considered poor based upon diagenetic variations.

*Source Rocks.* Source rocks are unknown resulting in a higher risk of hydrocarbon absence.

*Timing and Trapping Mechanisms.* Timing mechanisms are unknown. Stratigraphic pinch outs are anticipated but definite traps have not been identified.

*Resource Potential.* Petroleum exploration is placed at a very high risk of successfully developing hydrocarbons due to poor trap potential. Quantitative estimates of hydrocarbon accumulation could not be established.

## *Cody and Frontier Stratigraphic Play*

Deep oil and gas deposits are thought to accumulate in stratigraphic traps within the Cody and Frontier Formation marine shale and fine grained sandstone within the Torchlight and Peay members.

*Reservoirs.* Fine grained lithic sandstone appears to be evenly distributed throughout the play zone within the Cody and Frontier formations. Reservoir quality is expected to rapidly decrease with depth. Equivalent reservoirs located within structural traps are productive. It is expected that reservoir quality within deeper, off structural settings remain problematic.

*Source Rocks.* Cretaceous source rocks are present within the Mowry Shale.

*Timing Mechanisms.* Timing is unknown although trapping appears to have formed during primary hydrocarbon migration.

*Trapping Mechanisms.* Stratigraphic pinch out traps appear to be distributed throughout the play zone. Trapping mechanisms appear to be insignificant.

*Resource Potential.* Very high risks are associated with this play due to poor reservoir and trap potential. Significant hydrocarbon development is unlikely and there have not been any resource estimates completed.

### Shallow Tertiary-Upper Cretaceous Stratigraphic Play

Within the central parts of the Bighorn Basin, traps consist primarily of gas and minor oil within stratigraphic and combination traps.

*Reservoirs.* Reservoirs consist of sandstones, generally arkosic with good porosity and permeability at shallow depths. Principal reservoirs belong to the Fort Union, Lance, Meeteetsie, and Mesaverde Formations.

*Source Rocks.* Source rocks are associated with organic rich rocks belonging to Cretaceous aged formations. Gas appears to be of thermogenic origin with biogenic gas mixtures. Some vertical migration may have occurred.

*Timing Mechanisms.* Favorable hydrocarbon timing and migration occurred in traps that formed at the time of formational unit deposition.

*Trapping Mechanisms.* Stratigraphic traps formed by facies changes within locally formed alluvial channel sandstones. Traps are small and sometimes occur within combination stratigraphic and structural features. Seals are provided by fine grained rocks of upper Cretaceous age, Eocene, and Paleocene formations.

*Resource Potential.* Small accumulations are anticipated with a very high risk associated with occurrences of large accumulations.

Unconventional plays belong to geologic zones which produce significant amounts of natural gas, occasionally associated with oil, but often not.

### Basin Center Gas Play

Gas is trapped within the Bighorn Basin center within sandstone belonging to upper Cretaceous and Paleocene aged formations within the central basin zone. Accumulations of gas produces reservoir over-pressurization within Paleocene aged rocks. Cretaceous rocks may also be subjected to similar conditions but were not considered due to thin reservoir characteristics limiting reservoir volume. Production depths are limited to 12,000 feet to the base of the Mesaverde Formation.

*Reservoirs.* Reservoirs are lensed shaped, consisting of arkosic or lithic sandstone formed in alluvial settings with poor to moderate porosity and low permeability. Blanket type marine sandstone are present. The Fort Union, Lance, Meeteetsie, and Mesaverde formations contain reservoirs.

*Source Rocks.* Organic rich rock and coal appear to be the more common sources of gas deposits. Contributions may originate from the underlying Cretaceous formations. Gas origin appears to be thermogenic with some vertical migration possible.

*Timing Mechanisms.* Hydrocarbon generation and migration favors accumulation after trap formation. Gas generation appears to develop over-pressurized conditions within the reservoir.

*Trapping Mechanisms.* Accumulation appears related to regional stratigraphic traps caused by low reservoir permeability combined with active gas generation. Alluvial sandstone within local channelized deposits provides internal compartmentalization.

Trapping may be enhanced by local structures. Seals are provided by low reservoir permeability and by fine grained rocks within upper Cretaceous and Paleocene rocks. Ground water influx and hydrodynamic enhancement may contribute towards trapping mechanisms.

*Resource Potential.* Gas accumulation potential is good. Quantitative assessments have not been completed.

# References

Finn, T. M., 2010. Subsurface stratigraphic cross sections showing correlation of Cretaceous and Lower Tertiary rocks in the Bighorn Basin, Wyoming and Montana. Chapter 6 of Petroleum systems and geologic assessment of oil and gas in the Bighorn Basin Province, Wyoming and Montana. US Geological Survey Bighorn Basin Province Assessment Team, US Dept. of Interior.

Fox, J.E., Dalton, G.L., ___. Bighorn Basin Province (034). No published information.

Lageson, D. R., Spearing, D. R. 1988. Roadside Geology of Wyoming. Mountain Press Publishing Company, Missoula, MT.

Levorsen, A. I., 1967. Geology of Petroleum. W.H. Freeman and Company, San Francisco.

Szary, W.A. 2015. Introduction to Global Plate Tectonics II: North America, Alaska, The Appalachian Mountains, The Western US, Mexico, and the South American geologic histories. Earth2Energy Educational Publishing.

Thom, W.T. 1952. Tectonic team research – Key to social progress and world peace. Excerpted from Fanshawe, J. R. 2005. Bighorn Basin Tectonics. Southern Bighorn Basin, Wyoming. 7th Annual Field Conference Guidebook. Wyoming Geological Association Guidebook.

Wyoming State Geological Survey, 2016. Geological History of Wyoming, WSGS website publication.

# Index

Absaroka Mountains 21, 45, 50, 53, 56, 95, 106
Africa 7
Aleutian Islands 24
Alluvial 109
Alluvial fans 49, 50
Amsden limestone 31, 56, 102, 111
Anhydrite 32
Antarctica 6, 7
Antler arc 8, 9, 10, 27, 30
  Fore arc 12
  Foreland 12, 31
  Mountains 12, 16, 30
Arabia 7
Arctic 6, 37, 40
Atlantic Ocean 12
Australia 6, 7
Baja British Columbia 18, 21
Baltica 6
Basin & Range 23-26, 50, 51
Basalts 4
  Columbia River flows 25
  Flood 4
Batholiths 18, 40, 48, 80, 85
  Idaho 18, 20, 40
  Sierra Nevada 18, 20, 40
Beartooth Mountains 53, 89, 92, 104
Bighorn Basin 27, 38, 39, 44, 45, 53-56, 79, 87, 88-91, 94, 96, 99, 100, 103, 106, 107, 113, 114
  Tectonics 86
Bighorn dolomite 27, 56, 87, 102. 110
Bighorn Mountains 22, 47, 52, 53, 88
Bighorn River 53
Biogenic 113
Blanco fracture zone 24, 26
Blue Mountains 16, 17
Braided rivers 49
Bridger Mountains 53
British Columbia 26, 33, 35
Brooks Range 48
Cache Creek 16
  Inter-arc basin 16
Canada 4, 27. 41
  Alberta 35, 45
  British Columbia 35
  Manitoba 45
  Ontario 41
  Quebec 41
  Rockies  48
  Saskatchewan 35, 45, 46
Carbonate platform 27, 30, 32, 35
Cascades volcanic chain 23, 25-26
Casper 50
Challis arc 19, 21, 46
Chugwater Formation 34, 35, 54, 56, 89, 91, 96-98, 102, 105, 107
Clarks Fork River 53
Cloverly Formation 38, 56, 96, 102

Coastal plain 44, 107
Coast Ranges 16, 17
Cody Shale 43, 44, 89-91, 93-98, 103, 112
Colorado plateau 23, 25-26, 85
Cordilleran 82, 84
Cotton Creek Field 106
Cottonwood Field 104
Craton 3, 40, 42
　North American 3
Crazy Mountain Basin 53
Crevasse splay 44
Cross beds 44
Crow Mountain Formation 56, 102
Crustal extension 25, 51
Curtis Formation 105
Dakota region 42
Dakota Sandstone 56, 89-91, 102, 108, 109
Darby limestone 29
Darwin Formation 105, 111
Deltaic plain 109
Dikes 81
Dinwoody Formation 56, 102, 105, 107
Dryhead Agate 31
East Pacific Rise 18, 19, 22, 25
Elk Basin Anticline 89
Elk Butte 90
Embar anticline 96
Enos Creek Anticline 94, 95
Eolian 31, 107
Ervay Member 56, 106, 107
Estuarine 38, 39
Evaporites 33, 106
Farallon Plate 19, 21-23, 25, 46, 86, 87
Faults
　Detachment 89
　Reverse 47, 54
　Right lateral 21
　Slip faults 21, 25
　Thrust 3, 12, 14, 15, 20, 21, 47, 54, 82, 85
　Transform 25
Five Mile-Dobie Creek Field 103
Five Mile Trend 104
Flathead Formation 28, 56, 102, 110
Floodplain 38, 73, 76
Fluvial 38, 39, 44, 45
Foreland basins 22, 48, 88
Fort Union formation 49, 55, 89-91, 99, 113, 114
Fountain formation 30
Frontier Formation 43, 56, 89-91, 93, 94, 96-98, 101-104, 112
Fusion Shale 109
Gebo Anticline 97
Geosyncline 81, 85
Golden Eagle Anticline 96
Gooseberry Anticline 94
Goose Egg formation 32, 34, 35, 106, 108
Grand Teton Mountains 22. 47, 51
Granite Mountains 51
Grass Creek Anticline 94, 95

Great Basin 84
Great Lakes 40
Great Plains 22, 26, 84
Greenhorn formation 43
Grenville Mountains 5, 6
Greybull Formation 56, 102, 108
Greybull River 53
Guernsay limestone 30
Gulf of California 26
Gulf of Mexico 40, 45
Gypsum Springs Formation 36
Halite 32
Half Moon Anticline 92-94
Hamilton Dome 96
Hanna Basin 50
Heart Mountain Anticline 89
Hoback Range 86
Horse Creek Anticline 89, 90
Incised valley 39, 51
India 6, 7
Intrusions 4
  Igneous 4
Island arcs 3, 8, 11, 13-15, 17, 29, 30, 33, 35-37
  collision 3, 11-15, 17, 28, 30, 33, 35-37
Jackson Hole 51
Jefferson formation 27-28, 56, 102
Juan de Fuca Plate 24, 26
Karst 111
Kerogen 109
Kirby Creek Anticline 98
Klamath Mountains 16, 26
Kootenay Arc 11-13, 16
  Trench 20
Kula Plate 18, 19-24, 40, 46, 47
Lacustrine 38, 49, 82
Lake Creek Dome 98
Lakota Sandstone 56, 102, 108
Lance Formation 45, 95, 99, 113, 114
Lander formation 110
Laramide Mountains 19, 22, 39, 41, 45, 47, 51
Laramide Uplift 47-51, 53-57, 80, 82, 85-90, 100, 102, 103, 105
  Tectonics 86, 88
Laramie Basin 50
Laurentia 5-9
Lewis Formation 44
Lopoliths 80
Madison limestone 30-31, 56, 87, 91, 95, 99, 101-104, 110, 111
Marginal marine 44
McCloud Arc 14-16
Meeteetsie Anticline 91, 92,
Meeteetsie Formation 44, 90, 101, 102, 113, 114
Mendocino fracture zone 19, 22, 23, 25, 26
Mesaverde Formation 44, 56,90, 94, 95,99, 101-104, 113, 114
Mexico 9, 13, 18
Minnelusa Formation 108
Minnesota 80
Mississippi River 26
Missouri 42

Mohavia 5, 6
  Mohave desert 5
Morrison Formation 36, 109
Mowry Formation 42, 43, 101, 103, 109, 112
Muddy Sandstone 39, 56, 102, 103
Murphy Dome 98
Nebraska 42
Nevadan Arc 20
  Uplift 80
Neiber Dome 97
Neiber Thrust 97
North Sunshine Anticline 93, 94
Okanogan Arc 11, 12, 14, 16, 20
Olympic Mountains 24, 26
Oregon Basin Dome 89-91
Oregon Basin fault 45, 55
Overthrust Belt 48, 86
Owl Creek Fault 96
Owl Creek Mountains 52, 53
Paleo-topography 107-108
Paleo-valleys 38, 76
Pangea 12, 13, 29, 34
Pannotia 7, 9
Panthalassic Ocean 8
Pathfinder Uplift 32
Peay Member 112
Petroleum 57-99
  Formation 57
  Impervious rock 58, 60, 74
    Cap rock 58
    Roof rock 58
  Maturity 57
  Migration 57
  Pools 58-60, 62, 70, 71, 75, 78, 100
    Displaced 68
  Provinces 100
    Conventional Plays 100
      Basin Margin Anticlines 102
      Basin Margin Subthrust 100
      Deep Basin Structures 103
      Phosphoria Stratigraphic 106
      Sub Absaroka Play 104
    Hypothetical Plays 107
      Bighorn-Darby Wedge Edge Pinchout Play 110
      Cody & Frontier Stratigraphic Play 112
      Darwin-Amsden Sandstone Stratigraphic Play 111
      Flathead-Lander and Equivalent Sandstone Stratigraphic Play 110
      Greybull-Cloverly-Muddy Sandstone Stratigraphic Play 108
      Madison Limestone Stratigraphic Play 111
      Shallow Tertiary-Upper Cretaceous Stratigraphic Play 113
      Tensleep Paleo-topographic Play 107
    Unconventional Plays 114
      Basin Center Gas Play 114
Reservoir 57-61, 69, 72, 73, 100, 101, 103, 104
  Seeps 69, 70
  Traps 55, 58
    Combination 58-60, 69, 79, 113
    Fluid 78

Stratigraphic 55, 58, 59, 68, 73, 74, 100, 101, 109, 108, 112-114
  Carbonate rocks 76-78
    Bioherms 77
    Biostromes 77
    Fringing reefs 77
    Organic reefs 77
  Channel fill 76
  Lenticular 73, 74
  Lithofacies 74, 75
  Offshore sand bars 75
  Oxbow lakes & channels 76
  Primary 73, 74
  Secondary 73
  Shoestring sands 75
  Unconformities 63, 66-68
Structural 55, 57, 58, 59, 68, 69, 72, 100, 101, 112
  Anticlines 58-60, 64-66, 70
  Buried hills 66
  Competent beds 65
  Diapirs (salt plugs) 60, 65
  Faults 58-60, 68-71
  Folds 58-62, 64, 67, 68, 71, 72
    Asymmetrical 67
    Parallel 64
    Pattern shifting 62
  Fractures 60, 70, 72
  Incompetent beds 65
  Structural closure 60-61, 64, 71, 72, 109
  Structural relief 61, 63
Phosphate 32
Phosphoria formation 31, 55, 56, 101-103, 105, 106-108
Phosphorite 31
Pitchfork Anticline 91
Pinchout 43, 44, 110-113
Plates 12
  African 12
  European 12
  Juan de Fuca 24, 26
  North American 12, 20, 25, 26
  Pacific 12, 15, 16, 20, 24, 26
    Ancestral 15, 21, 40
Playa lakes 49, 82
Powder River Basin 47, 50, 108
Precambrian Era 3, 5, 81
  Rifts 81, 85
Prodeltaic environment 43, 44, 76
Proterozoic 7
Pryor Mountains 53, 88
Rattlesnake Creek Field 104
Rattlesnake Mountain 87, 90, 92
Rawhide Anticline 91
Red Canyon fold 97
Red Peak Formation 34, 35
Rifts 6, 9, 20, 25, 26
Roberts Mountain Thrust 10, 12, 30
Rocky Mountains 12, 22, 45, 47, 83, 85
  Ancestral 29, 30, 32
  Foreland 100

Front Range 81
Rodinia 4-7, 9
Rose Dome 97
Rhyolite ash 51
Salt River Range 86
San Andreas Fault 24-26
Sevier Mountain Uplift 18, 21, 39-41, 86
  Tectonics 86, 87
Shearing 15
  Left lateral 15
Shirley basin 50
Shoshone River 53
Siberia 6
Sierra Mountains 13, 17, 18, 21, 26, 46
Skelton Dome 94, 95
Snake River Plain 24, 26, 51
Snake River Range 86
Sonomian Arc 13, 14, 16
  Mountain belt 15, 16
South Africa 6
South America 6
South China 5, 6
South Oregon basin 92, 93
South Sunshine Anticline 93, 94
Spearfish Formation 33
Spring Creek Anticline 91
Stream channels 44, 49, 73, 75
Sublette Range 86
Sundance Formation 36
Sweetwater Hills 51
Tatman Formation 49
Teapot sandstone 44
Tectonics 6
  Accretion 6, 11, 12, 15, 19, 21, 23, 34
  Suture zones 6, 20
Tensleep sandstone 31, 55, 56, 87, 96, 101-105, 107, 108
Texas 42, 45
Thermogenic 113, 114
Thermopolis, town 97
Thermopolis Shale 38, 43, 103, 109
Thick skinned tectonics 87, 88
Thin skinned tectonics 86
Tidal channels 106
Tidal flats 38, 76, 106, 107
Torchlight Member 112
Transform faults 19
Transgressive-regressive cycles 44
Tropical seas 27
Tunip Range 86
Unconformity 43, 108, 110
volcanic chains 3, 21, 47
  arcs 6, 9, 11, 14, 16, 17, 20, 40, 48
Wallowa Mountains 16
Warm Springs 98
Wasatch Formation 55, 92, 94, 96
Washaki Mountains 53
West Oregon basin 92
Western Interior Seaway 18, 19, 21, 35, 36, 40, 41, 42, 44-47, 50

Western US 3, 5, 6, 26, 36, 51
  Arizona 12, 16, 19, 35, 36, 46
  California 5, 6, 9, 12, 15, 16, 20, 21, 25, 26, 40, 46, 86
  Colorado 12, 25, 29, 36, 37, 40, 45
  Idaho 6, 12, 15, 16, 21, 22, 25, 26, 33, 39, 46, 51, 81, 107
  Montana 6, 12, 21, 22, 35, 40, 41, 46-48, 51, 81, 84, 85
  Nevada 6, 16, 17, 20, 50, 86
  New Mexico 12, 25, 40-42, 45
  North Dakota 35
  Oregon 5, 6, 12, 13, 16, 20, 21, 25, 40, 46
  Utah 5, 12, 21, 33, 36, 41, 42, 46, 48, 50, 55, 86
  Washington 5, 6, 11-13, 20, 24-26, 39
Wild Horse Butte 98
Willwood formation 49, 55, 89, 93, 96
Wind River Basin 100
Wind River Indian Reservation 96, 97
Wind River Range 47, 50, 51
Wisconsin 40, 42
Worland Field 104
Wrangellia 16, 17, 20
  Uplift 16
Wyoming 4, 21, 22, 27, 28, 30, 32, 34, 36, 37, 39, 40, 45-51, 54, 79, 86
  Central 28, 30, 35, 37, 42, 43, 46, 47, 53, 56
  Cheyenne Belt 4
  Eastern 8, 28, 30, 32, 35, 37, 42, 43, 45, 46, 56
  Northern 35
  Northwest 50
  Plate 4
  Southeastern 32
  Southern 37
  Southwestern 28, 47-50, 100
  Western 8, 21, 28-31,36, 37, 41, 42, 46, 48, 55, 57, 86, 102, 107
Yellowstone Hot Spot 24-26
  Volcano 24
Yellowstone region 82
  Boundary 105
Yellowstone River 53
Yukon Territory 27

All book titles are intended for those interested in earth sciences at the secondary school, community college, and first year undergraduate level of study. Technical terms are defined in the text where appropriate.

**Introduction to Global Plate Tectonics I: Plate Tectonics Theory, Paleogeography, and the Ocean Basins. By William A. Szary. Earth2Energy Educational Publishing, Tampa, FL.** This book is intended for those interested in general geology. Contents include many images to help the reader understand the underlying principles used to explain plate tectonics. The book summarizes plate tectonic theory by explaining the various driving forces behind continental drift theory. Text is presented in plain language for those interested in learning about the basic principles geologists use to explain the positioning of continents around the globe. Some technical terms are used, but are defined as they are presented. Chapter I **Plate Tectonics Theory** presents geologic models redrawn from the National Geographic Society, and by diagrams prepared by the United States Geological Survey. The text helps to elaborate on the image captions with more detail. Chapter II **Paleogeography** presents the principles of continental drift through global map reconstructions showing the assemblies of the continents through geologic time from the Precambrian Era Rodinia supercontinent through Gondwana, and the breakup of Pangea. The globe map series were produced by Dr. C.R. Scotese of the PaleoMap Project. Maps were used with permission from Dr. Scotese. Chapter III presents the geologic history of the major **Ocean Basins** formed by continental drift and collision throughout geologic time. A series of maps produced by the National Geographic Society, reconstruction drawings published in technical journals including the Deep Sea Drilling Project, and globe maps prepared by Robert Hall of the SE Asia Research Group are used to present graphics which accompany the narrative. **Purchase online through Createspace eStore at** https://www.createspace.com/4931771. **Price: $29.99.**

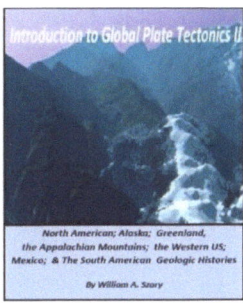

**Introduction to Global Plate Tectonics II: North America, Alaska, The Appalachian Mountains, the Western US, Mexico, & South American Geologic Histories. By William A. Szary. Earth2Energy Educational Publishing, Tampa, FL.** The second part of a five part series covering the subject of plate tectonics, paleogeography, and the drifting and buildout of continents. Part II covers the development of the North American basement to the present with a peek into future continental arrangements. Chapters on Alaska, the Appalachian Mountains, Western US, Mexico, and South American geologic histories are included. **Purchase online through Createspace eStore at** https://www.createspace.com/4950697. **Price: $49.99.**

**Introduction to Global Plate Tectonics III: Europe, Russia, Mongolia, China, Korea, & Indochina Geologic Histories. By William A. Szary. Earth2Energy Educational Publishing, Tampa, FL.** The third in a five part series covering the development of the European basement to the present day. Chapters on Russia, Mongolia, China, Korea, and Indochina geologic histories are included. **Purchase online through Createspace eStore at** https://www.createspace**.com/4957439. Price: $39.99.**

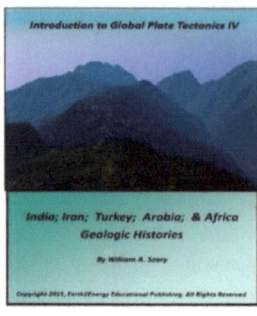

**Introduction to Global Plate Tectonics IV: India, Iran, Turkey, Arabia, and Africa Geologic Histories. By William A. Szary. Earth2Energy Educational Publishing, Tampa, FL.** Part four covers the development of the Indian, Iranian, Turkish, Arabian, and African basement to the present day. This book approaches the subject matter in a more technical forum.
**Purchase online through Createspace eStore at** https://www.createspace.com/5166539**. Price: $49.99.**

**Introduction to Global Plate Tectonics V: Australia & Antarctica Tectonic Evolution. By William A. Szary. Earth2Energy Educational Publishing, Tampa, FL.** The last of the five part series covering the development of the Australian and Antarctica Precambrian Era basement continuing to present day. This book focuses on the more technical aspects of plate tectonic evolution.
**Purchase online through Createspace eStore at** https://www.createspace.com/5189356**. Price: $19.99.**

### *Introduction to Geomorphology Series*

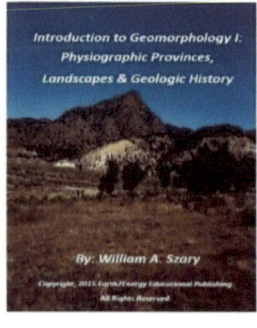

*Introduction to Geomorphology I: Physiographic Provinces, Landscapes, & Geologic History. By William A. Szary. Earth2Energy Educational Publishing, Tampa, FL.* Introduction to Geomorphology I reviews the geologic history behind each physiographic province providing typical and atypical photographic representations for each recognized province in the continental US.
**Purchase online through Createspace eStore at** https://www.createspace.com/5223816**. Price: $39.99.**

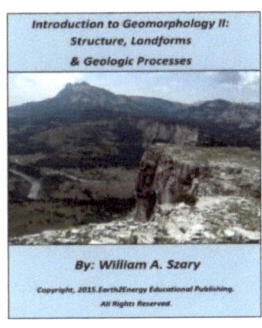

Introduction to Geomorphology II: Structure, Landforms, & Geologic Processes. By William A. Szary. Earth2Energy Educational Publishing, Tampa, FL. Book II continues with expanding the description of geomorphic provinces describing landscapes in the context of geologic structure, landforms, and basic principles which shape landforms and landscapes. Many photographs are presented in this book covering constructive, destructive, mass wasting, fluvial, glacial, and coastal processes. **Purchase online through Createspace eStore at https://www.createspace.com/5369007. Price:** $49.99.

## Florida Geology Series

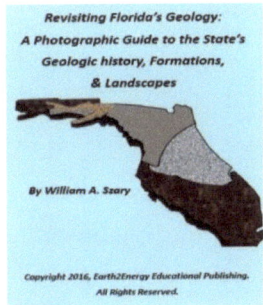

**Revisiting Florida's Geology** presents a summary of the state's early geologic history during the Paleozoic and Mesozoic Eras including a discussion on geologic formations, structures, and tectonic processes forming the basement complex. The Cenozoic history is presented in the context of the uppermost limestone and sedimentary rock formations. Various types of landscapes are presented using selected county geologic maps to show which formations are responsible for producing flatlands, gently to moderately sloping hills, steep hills, valleys, and karstic processes. **Purchase online through Createspace eStore at https://www.createspace.com/5931881. Price:** $59.99.

## Other Titles

The Sierra Nevada Foothills presents an in depth analyses of the characteristics associated with a subduction zone complex based on a field study completed in the Sierra Nevada Foothills Terrane near Valley Springs, California during 2000. A proposed geologic map was compiled based on observations completed on a limited number of rock outcrop exposures, and through the use of satellite imagery obtained in 2005. A series of plate tectonic models are presented offering theories on accretion rates, erosion rates, plate collision rotation, etc. The book is intended for advanced geology students at the third to fourth level undergraduate and graduate studies level in the geological sciences. **Purchase online through Createspace eStore at https://www.createspace.com/6231576. Price:** $59.99.

*Earth2Energy Educational Publishing*
17749 Jamestown Way #F, Lutz FL 33558

www.ingramcontent.com/pod-product-compliance
Lightning Source LLC
Chambersburg PA
CBHW040807200526
45159CB00022B/41